SpringerBriefs in Statistics

JSS Research Series in Statistics

The current research of statistics in Japan has expanded in several directions in line with recent trends in academic activities in the area of statistics and statistical sciences over the globe. The core of these research activities in statistics in Japan has been the Japan Statistical Society (JSS). This society, the oldest and largest academic organization for statistics in Japan, was founded in 1931 by a handful of pioneer statisticians and economists and now has a history of about 80 years. Many distinguished scholars have been members, including the influential statistician Hirotugu Akaike, who was a past president of JSS, and the notable mathematician Kiyosi Itô, who was an earlier member of the Institute of Statistical Mathematics (ISM), which has been a closely related organization since the establishment of ISM. The society has two academic journals: the Journal of the Japan Statistical Society (English Series) and the Journal of the Japan Statistical Society (Japanese Series). The membership of JSS consists of researchers, teachers, and professional statisticians in many different fields including mathematics, statistics, engineering, medical sciences, government statistics, economics, business, psychology, education, and many other natural, biological, and social sciences. The JSS Series of Statistics aims to publish recent results of current research activities in the areas of statistics and statistical sciences in Japan that otherwise would not be available in English; they are complementary to the two JSS academic journals, both English and Japanese. Because the scope of a research paper in academic journals inevitably has become narrowly focused and condensed in recent years, this series is intended to fill the gap between academic research activities and the form of a single academic paper. The series will be of great interest to a wide audience of researchers, teachers, professional statisticians, and graduate students in many countries who are interested in statistics and statistical sciences, in statistical theory, and in various areas of statistical applications.

More information about this subseries at http://www.springer.com/series/13497

Nobuaki Hoshino · Shuhei Mano · Takaaki Shimura
Editors

Pioneering Works on Extreme Value Theory

In Honor of Masaaki Sibuya

Editors
Nobuaki Hoshino
School of Economics
Kanazawa University
Kanazawa, Ishikawa, Japan

Shuhei Mano
The Institute of Statistical Mathematics
Tachikawa, Tokyo, Japan

Takaaki Shimura
The Institute of Statistical Mathematics
Tachikawa, Tokyo, Japan

ISSN 2191-544X ISSN 2191-5458 (electronic)
SpringerBriefs in Statistics
ISSN 2364-0057 ISSN 2364-0065 (electronic)
JSS Research Series in Statistics
ISBN 978-981-16-0767-7 ISBN 978-981-16-0768-4 (eBook)
https://doi.org/10.1007/978-981-16-0768-4

This Springer imprint is published by the registered company Springer Nature Singapore Pte Ltd.
The registered company address is: 152 Beach Road, #21-01/04 Gateway East, Singapore 189721, Singapore

Preface

Amid the COVID-19 pandemic, we all recognize that an extreme event can be fatal in life. Extreme value theory is undeniably vital. This book aims to present its frontier to an audience who is never satisfied with the pure theory. Refereed articles, ranging from theory to practice, have been compiled into this volume dedicated to Professor Masaaki Sibuya.

He is among the key founders of the Japanese Society of Applied Statistics and has always been advancing theory for practice. His pioneering attitude can be discerned in his seminal paper on the bivariate extreme value distribution published in 1960. He conceptualized the dependence of extreme events, the negligence of which results in a catastrophe historically. Extreme events do simultaneously occur. For example, bankruptcy is usually regarded as an extreme event. However, we observe serial bankruptcies under the current pandemic, not to mention the global financial crisis in 2008. Floods, earthquakes, and tsunamis, the geographical dependence of these extremal issues have been crucial in the fight against natural disasters. Professor Sibuya induced the decisive factor of practice antecedently.

Initiating this commemorative volume, his other contribution against a tsunami is presented in Chapter 1. Premodern tsunamis were recorded in various ways such as remains, a monument, and a diary. These historical records, however, would neglect small tsunamis. In other words, an unknown infimum should exist in tsunami records. The supremum magnitude of a tsunami is obviously unknown. The generalized beta distribution has the parameters of the infimum and the supremum, which can describe tsunami records. However, the unknown support invalidates the regular asymptotics of the maximum likelihood estimation. Professors Mano and Sibuya proposed a new method to estimate the parameters of the generalized beta distribution notwithstanding the irregularity. The editors would also remind readers of the presented application to the Japanese historical data of tsunamis.

The dependence of events can be flexibly modeled through a copula. In 2017, Professor Sibuya proposed the empirical beta copula, a nice smoother of the empirical copula, jointly with Professors Segers and Tsukahara. Chapter 2 confirms that the empirical beta copula provides expedient resampling procedures. Professors Kiriliouk, Segers, and Tsukahara showed the asymptotic equivalence of several bootstrapped processes related to the empirical and empirical beta copulas. Moreover, they clarified the finite sample properties of these resampling schemes by extensive simulations.

Chapter 3 employs a copula and extreme value theory in an interesting way. A binary regression model, such as logistic regression, explains the probability of an event by mapping a real number to a unit interval [0,1] using a cumulative distribution function. In this case, there exists a limiting argument to let the probability of an event be very small so that the extreme value distributions appear through the min-stability, depending on the distribution function used for mapping. Professor Sei noted this fact in his previous study; the distribution function was currently generalized to include multi-dimensional functions. He obtained the corresponding limit, exploiting the max-stability of a copula.

The estimation of the support is revisited in Chapter 4, where Doctor Ozeki and Professor Doksum consider a semiparametric regression model. In their argument, the conditional distribution of a response variable given a covariate vector is unknown as well as the upper boundary of its support. However, the boundary is assumed to be a function, say g, of the linear combination of covariates. The coefficients of this combination are boundary regression parameters to be estimated. After Le Cam's asymptotic theory for the irregular estimation was reshaped to include semiparametric models, they constructed a consistent and efficient estimator for the known function g. Also for the unknown g in a parametric class, a consistent estimator was obtained.

The remaining two chapters focus on the hydrological practice, in particular, the analysis of extreme rainfalls for the flood control of Japanese rivers. They are short and steep, owing to the mountainous land. These characters have fostered the specific adoption of statistics.

The frequency analysis of extremes usually discards non-extreme observations. However, the definition of extreme events varies. For time-series observations, the whole duration is first partitioned into short periods (blocks), and extremes are extracted as samples in each block. These extremes may be limited to the maximum or all instances over a threshold. The latter carries more information but may be sensitive to the selection of a threshold. Professor Tanaka, in Chapter 5, performed the sensitivity analysis of this phenomenon based on past observations since 1876 and future simulation data. The necessity of sensitivity analysis is described from a more general viewpoint in Chapter 6. Professor Takara reviewed the past development of hydrologic frequency analysis in Japan, providing a valuable summary of his long experience. Diverse goodness-of-fit criteria and resampling schemes have been compared to yield a robust conclusion.

Chapter 1 was prepared exclusively for this volume. Other chapters are based on presentations at the Pioneering Workshop on Extreme Value and Distribution Theories in Honor of Professor Masaaki Sibuya held at the Institute of Statistical Mathematics (ISM), Tokyo, Japan from March 21–23, 2019. This workshop was sponsored by Kanazawa University and the ISM and co-sponsored by the Advanced Innovation powered by the Mathematics Platform and Japanese Society of Applied Statistics. This workshop and the editing process for this volume and one more volume entitled "Pioneering Works on Distribution Theory: In Honor of Masaaki

Sibuya" were financially supported by the KAKENHI grant (18H00835). The editors appreciate the significant support from these sponsors. Also, their appreciations go to people who cooperated in bringing this work to reality. In particular, the reliability and accuracy of the information herein were made possible due to the generosity of many esteemed anonymous reviewers.

Last, but certainly not least, the editors would like to express their sincere gratitude to Professor Sibuya for his decades of mentorship to each of the editors.

Tokyo, Japan
December 2020

Nobuaki Hoshino
Shuhei Mano
Takaaki Shimura

Contents

Chapter 1
Parameter Estimation of Generalized Beta Distributions and Its Application to a Historical Tsunami Magnitude Dataset

Shuhei Mano and Masaaki Sibuya

Abstract For estimating the generalized (four-parameter) beta distributions, Nagatsuka-Balakrishnan-Kamakura transformation and Hall and Wang's empirical Bayesian likelihood are applied to the shape parameter and the location-scale parameter, respectively. With this procedure and a smoothing, a new estimator of parameters is proposed in this paper. Some non-normal limit distributions of the estimators of the location-scale parameter are discussed, and the performance of the proposed estimator is evaluated with the estimator by Hall and Wang's method and that by Kachiashvili et al.'s method. The proposed method is applied to a historical dataset of tsunami magnitude scales.

Keywords Four-parameter beta distributions · Hall and Wang's empirical Bayesian likelihood · Location-scale parameters · Nagatsuka-Balakrishnan-Kamakura transformation · Non-regular estimation · Non-standard beta distributions · Non-standard estimation · Reverse Weibull distributions · Smoothing · Tsunami scale

1.1 Introduction

Let the pdf of the three-parameter gamma distribution with a threshold be denoted by

$$f_1(x; \theta, c) = g_1(x - c; \theta), \quad c < x < \infty,$$

S. Mano (✉)
The Institute of Statistical Mathematics, 10-3 Midori-cho, Tachikawa, Tokyo 190-8562, Japan
e-mail: smano@ism.ac.jp

M. Sibuya
Keio University, 3-14-1 Hiyoshi, Kohoku-ku, Yokohama, Kanagawa 223-8522, Japan
e-mail: sibuyam@1986.jukuin.keio.ac.jp

N. Hoshino et al. (eds.), *Pioneering Works on Extreme Value Theory*, JSS Research Series in Statistics,
https://doi.org/10.1007/978-981-16-0768-4_1

where $g_1(x;\theta),\ \theta=(a,h),\ a,h>0$ is the standard gamma distribution with the shape parameter a and scale parameter h. Since the support depends on the threshold c, the classical asymptotic properties of the maximum likelihood estimation fail to apply in this case [19]. In order to solve the non-regular estimation of (θ,c), Nagatsuka, Balakrishnan, and Kamakura [22] used the order statistics of a sample from $f_1(x;\theta,c)$ for eliminating c from the estimation of θ. They were mainly concerned with the estimation of θ, and a moment estimator was used for c. They also discussed parameter estimation of the three-parameter Weibull distribution in a similar way [21].

On the other hand, Hall and Wang [14] examined the estimation of the threshold c of the pdf of the form

$$f_2(x;\theta,c)=(x-c)^{\gamma-1}g_2(x;\theta,c),\quad c<x<\infty, \tag{1.1}$$

where $\gamma>0$, $g_2(x;\theta,c)\to g_0>0$, $(x\downarrow c)$ for a constant g_0, and θ denotes a vector of parameters other than c. The form includes the three-parameter gamma and Weibull distributions. For the joint estimation of (γ,c), the pdf is unbounded and the global maximum does not exist; if $c\ll x_{(1)}$, where $x_{(1)}$ is the smallest observation, the pdf diverges as γ goes to infinity. However, when $\gamma>1$, there exists a local maximum, which is used as a consistent estimator of c [26]. For $\gamma>2$, although the usual regularity conditions for the maximum likelihood estimation are not satisfied, the less-stringent conditions of Le Cam [18] are satisfied for the local maximum likelihood estimator. The estimator has the same asymptotic properties as in usual regular cases; the asymptotic normality holds with the usual convergence rate of $O_p(n^{-1/2})$ [28], where n is the sample size, and the variance asymptotically equal to the Cramér-Rao lower bounds. For $\gamma=2$, the asymptotic normality holds, but the convergence rate is $O_p\{(n\log n)^{-1/2}\}$ [4, 28]. For $1<\gamma<2$, the local maximum likelihood estimators have a non-normal limit distribution with a convergence rate of $O_p(n^{-1/\gamma})$ [29]. In contrast to these cases, for $\gamma\le 1$, the local maximum does not exist and the maximum likelihood estimation fails to produce a solution other than $\hat{c}=x_{(1)}$, which is clearly biased. To overcome the non-availability of the maximum likelihood estimator, Hall and Wang [14] proposed the maximum Bayesian likelihood estimation with an empirical prior

$$p_1(c)=\frac{x_{(1)}-c}{x_{(2)}-c}, \tag{1.2}$$

using the first two order statistics $x_{(1)}$ and $x_{(2)}$ of a sample from $f_2(x;\theta,c)$. The prior is multiplied to the likelihood $\prod_{i=1}^n f_2(x_i;\hat{\theta}_c,c)$, where $\hat{\theta}_c$ is the maximum profile-likelihood estimate, and the product, called *the Bayesian likelihood*, is maximized with respect to c.

The ideas are extended to the family of finite interval densities

$$f(x;\theta,c,d)=\frac{1}{d-c}\,g\left(\frac{x-c}{d-c};\theta\right),\quad c<x<d, \tag{1.3}$$

where $g(y;\theta)$, $0 < y < 1$ is a pdf independent of the location-scale parameter (c, d). We are interested in the beta distribution

$$g(y;\theta) = \frac{1}{B(a,b)}\, y^{a-1}(1-y)^{b-1}, \quad 0 < y < 1, \tag{1.4}$$

with the shape parameter $\theta = (a, b)$, $a > 0$, $b > 0$. The family of densities

$$\begin{aligned} f(x;\theta,c,d) &= \frac{1}{d-c}\, g\left(\frac{x-c}{d-c};\theta\right) \\ &= \frac{1}{B(a,b)}\left(\frac{(x-c)^{a-1}(d-x)^{b-1}}{(d-c)^{a+b-1}}\right), \quad c < x < d, \end{aligned} \tag{1.5}$$

with $c > -\infty$ and $d < \infty$, is called *generalized beta distributions* (or *four-parameter beta distributions*). Johnson, Kotz and Balakrishnan [15] called (1.5) beta distributions, while they called (1.4) the standard form. Since the beta distribution is a commonly used model of distributions over a finite interval, the parameter estimation of generalized beta distributions has been discussed by many authors. Carnahan [6] discussed the local maximum likelihood estimation for the restricted case of $\min(a, b) > 2$, where the conditions of Le Cam [18] are satisfied for the local maximum likelihood estimators. Based on a numerical study, he showed that it is only for very large samples that the bias becomes small and the Cramér–Rao bound becomes a good approximation of the variance. Cheng and Iles [7] suggested the use of a corrected likelihood, which require c being estimated by $x_{(1)}$ if $a < 1$ and d by $x_{(n)}$ if $b < 1$.

Wang [27] proposed the use of two priors:

$$p_1(c), \;\text{ and }\; p_2(d) = \frac{d - x_{(n)}}{d - x_{(n-1)}} \tag{1.6}$$

and discussed parameter estimation based on the Bayesian likelihood. Previous studies on the generalized beta distributions have discussed joint estimation of the four parameters. Based on a numerical study, Wang [27] observed that (i) the correlation between the estimators of c and d are weak, and (ii) the estimators of c are more correlated to the estimator of a than to that of b, and the estimator of d is more correlated to the estimator of b than to that of a.

As we have seen so far, the shape parameter θ and the location-scale parameter (c, d) have quite different properties: θ determines regularity of the estimation problem, while (c, d) does not. In the joint estimation, we can expect that the information of θ will help in the estimation of (c, d) and vice versa, but the sharing of information could cause biases to the estimates. In the joint estimation, the likelihood is unbounded and the global maximum does not exist. Nevertheless, once θ is fixed, the estimation of (c, d) is more straightforward; the likelihood is bounded and the global maximum likelihood estimator is useful. Given these reasons, the separate estimation of θ and (c, d) is worth investigation.

In the context of parameter estimation of the generalized beta distribution, the idea of separating the estimations of θ and (c, d) was recently employed by Kachiashvili et al. [16, 17]. Given (c, d), a moment estimator of θ can be computed, while given θ, an unbiased estimator of (c, d) can be computed. Kachiashvili et al. proposed joint estimation of θ and (c, d) by iterating these two steps. Their method demands few computations, because these two estimators have simple expressions. However, they reported that their estimator is significantly biased [17].

In Sect. 1.2, we combine the above-mentioned two ideas: the estimation of shape parameter θ is separated from the estimation of the location-scale parameter (c, d) by using the transformation proposed by [21, 22], which will be called the NagaBalaKama transform, and then (c, d) is estimated based on the Bayesian likelihood with the empirical prior (1.6) [14, 27]. With this procedure and a smoothing, a new estimator of parameters of generalized beta distributions with pdf (1.5) is proposed. In Sect. 1.3, non-normal limit distributions of the estimators of the location-scale parameter are discussed, the performance of the estimator is evaluated numerically, and the estimator is applied to a historical dataset of tsunami magnitudes.

1.2 Estimation Methods

1.2.1 NagaBalaKama Transform for Estimating Shape Parameters

Let us assume the location-scale parameter (c, d) is known and fixed. Let $\mathbf{X} = (X_1, \ldots, X_n)$ be an iid sample from the generalized beta distribution of pdf (1.3), i.e., $\mathbf{Y} := (\mathbf{X} - c)/(d - c)$ is an iid sample from the beta distribution of pdf (1.4). Define $W_1, \ldots, W_n$ by

$$W_i := \frac{X_{(i)} - X_{(1)}}{X_{(n)} - X_{(1)}} = \frac{Y_{(i)} - Y_{(1)}}{Y_{(n)} - Y_{(1)}}, \quad 0 = W_1 < W_2 < \cdots < W_{n-1} < W_n = 1,$$

where $X_{(1)} < \cdots < X_{(n)}$ and $Y_{(1)} < \cdots < Y_{(n)}$ are the order statistics of $\mathbf{X}$ and $\mathbf{Y}$, respectively. Note that $(W_2, \ldots, W_{n-1})$ is independent of (c, d), because $(Y_{(1)}, \ldots, Y_{(n)})$ is independent of (c, d). Moreover, $(Y_{(1)}, \ldots, Y_{(n)})$ is the type-II censored sample from the pdf (1.4). The joint distribution of $(W_2, \ldots, W_n, Y_{(1)}, Y_{(n)})$ is given as

$$\begin{aligned}
&P(W_2 < w_2, \ldots, W_{n-1} < w_{n-1}, Y_{(1)} < u, Y_{(n)} < v) \\
&= P(Y_{(2)} < u + w_2(v - u), \ldots, Y_{(n-1)} < u + w_{n-1}(v - u), Y_{(1)} < u, Y_{(n)} < v) \\
&= n! \prod_{i=1}^{n} G(u + w_i(v - u); \theta),
\end{aligned}$$

where in the last equation we used the fact that $\mathbf{Y}$ is an iid sample from the pdf (1.4) and

$$G(u;\theta) = \int_0^u g(y;\theta)dy.$$

The joint density is given as

$$n!(v-u)^{n-2}\prod_{i=1}^{n} g(u + w_i(v-u);\theta).$$

Integrating out (u, v) with $u < v$, we obtain the likelihood of $(w_2, \ldots, w_{n-1})$:

$$l_{NBK}(w_2, ..., w_{n-1};\theta) = n!\int_0^1\int_0^v (v-u)^{n-2}\left(\prod_{i=1}^{n} g(u+(v-u)w_i;\theta)\right)dudv, \tag{1.7}$$

where $w_1 = 0$ and $w_n = 1$. We shall call the map $(X_{(1)}, \ldots, X_{(n)}) \longrightarrow (W_1, \ldots, W_n)$ as *the NagaBalaKama transform*. The expression (1.7) makes it possible to estimate the shape parameter $\theta = (a, b)$ apart from (c, d), based on $\mathbf{W}$ or $\mathbf{X}$. The separations of the estimation of the shape parameter from the estimation of the location-scale parameter were demonstrated for the three-parameter gamma distribution [22] and the three-parameter Weibull distribution [21]. Moreover, it should be stressed that the separation is applicable to generic location-scale families, because, in the derivation of (1.7), we did not use specific properties of the (generalized) beta distributions.

Given the estimate of θ, say $\hat{\theta} = (\hat{a}, \hat{b})$, a simple estimate of (c, d) is the moment estimate. Note that

$$E(X) = c + (d-c)E(Y) \quad \text{and} \quad SD(X) = (d-c)SD(Y),$$

where E denotes expectation and SD denotes standard deviation. Replacing $E(X)$ and $SD(X)$ with the sample mean and standard deviation $\widehat{\mu_X}$ and $\widehat{\sigma_X}$, we have the moment estimators

$$\hat{c} = \widehat{\mu_X} - \widehat{\sigma_X}\sqrt{\frac{\hat{a}}{\hat{b}}(\hat{a}+\hat{b}+1)}, \quad \text{and} \quad \hat{d} = \widehat{\mu_X} + \widehat{\sigma_X}\sqrt{\frac{\hat{b}}{\hat{a}}(\hat{a}+\hat{b}+1)}.$$

An apparent problem with the moment estimate is that the interval $(\hat{c}, \hat{d})$ does not always cover the range of a sample $(x_{(1)}, x_{(n)})$. Hence, the moment estimators are defective.

1.2.2 Hall and Wang's Empirical Prior for Estimating Location-Scale Parameters

Hall and Wang [14] explained the nice properties of the empirical prior $p_1(c)$, (1.2);

1. $p_1(c) > 0$ implies $c < x_{(1)}$.
2. $p_1(c) \approx 1$ implies $c \ll x_{(1)}$ or $x_{(1)} \approx x_{(2)}$ and the prior is non-informative.
3. When $x_{(1)} \ll x_{(2)}$, $p_1(c)$ rises slowly from 0 to 1 as c decreases, indicating that c is not near $x_{(1)}$.
4. When $x_{(1)} \approx x_{(2)}$, $p_1(c)$ rises sharply to 1 as c decreases, indicating that c is close to $x_{(1)}$.

Hence, $\log p_1(c)$ works as a penalty to avoid c coming close to $x_{(1)}$ and keeps the information on c.

In the family of generalized beta distributions, if d is known, then the pdf of the family reduces to the form of (1.1), and our concern is the left tail. The right tail behaves similarly because of the symmetry of the density (1.5) under $(x - c) \mapsto (d - x)$ by exchanging a and b. Further, if $(X_{(1)}, \ldots, X_{(n)})$ is the order statistics of a random sample of a pdf, $(X_{(1)}, X_{(2)})$ and $(X_{(n-1)}, X_{(n)})$ are asymptotically independent (see Lemma 1.1). Hence, Hall and Wang's empirical Bayes estimation is applicable to the generalized beta distributions with the empirical priors (1.6) [27]. The Bayesian log-likelihood to be minimized is

$$lkh(x; \theta, c, d) = \log(p_1(c)) + \log(p_2(d)) - n\log(B(a, b)) - n(a + b - 1)\log(d - c) + (a - 1)\sum_{i=1}^{n} \log(x_{(i)} - c) + (b - 1)\sum_{i=1}^{n} \log(d - x_{(i)}). \quad (1.8)$$

Wang [27] considered the joint estimation of the four parameters. She proposed the use of the stepwise ascent. It starts with some rough estimates of the parameters, followed by improving one parameter at a time while keeping the other parameters fixed, and increasing the posterior density at each step by using the Newton–Raphson method. The steps will eventually converge to the posterior local mode, and the resulting parameters will give the local maximum likelihood estimates. Although she did not mention it, the parameter space should be restricted, since the likelihood is unbounded and the global maximum does not exist; if $c \ll x_{(1)}$ and $d \gg x_{(n)}$, the likelihood diverges as either of a or b goes to infinity. For small n, the search tends to seek the global maximum which produces nonsense estimates. Based on numerical studies, Carnahan [6] reported this tendency for the case of $\min(a, b) > 2$. See Table IV of [6].

1.2.3 Kachiashvili's Iteration for Joint Estimation

At the end of Sect. 1.2.1, we pointed out that the moment estimators of the location-scale parameter (c, d) given the shape parameter θ is not useful. In contrast, the moment estimator of θ given (c, d) is useful. Kachiashvili and Prangishvili [16] pointed out that the iterative use of the moment estimator and an unbiased estimator of (c, d) given θ provides a joint estimate of θ and (c, d). Kachiashvili and Melikdzhanjan [17] investigated the performance of the method based on numerical studies.

Kachiashvili et al.'s method consists of the following steps.

1. $c = x_{(1)}$ and $d = x_{(n)}$.
2. Compute
$$a = \frac{\bar{x} - c}{d - c}\{(\bar{x} - c)(d - \bar{x})s^{-1} - 1\}, \quad b = \frac{d - \bar{x}}{d - c}\{(\bar{x} - c)(d - \bar{x})s^{-1} - 1\},$$
if $(\bar{x} - c)(d - \bar{x}) > s$, where $\bar{x} = \sum x_i$, $s = \sum (x_i - \mu)^2/(n - 1)$.
3. Compute
$$c = x_{(1)} - (x_{(n)} - x_{(1)})\frac{h_{(1)}}{1 - h_{(1)} - h_{(n)}}, \quad d = x_{(n)} + (x_{(n)} - x_{(1)})\frac{h_{(n)}}{1 - h_{(1)} - h_{(n)}},$$
where $h_{(1)} = \int_0^1 \{1 - G(y; \theta)\}^n dy$, $h_{(n)} = \int_0^1 \{G(y; \theta)\}^n dy$, and $G(y; \theta)$ is the distribution function of the beta distribution.
4. Stop if the Kolmogorov–Smirnov statistic $\sup_x |F_n(x) - F(x; \theta, c, d)|$ takes the minimum, where $F_n(x)$ is the empirical distribution function of the generalized beta distribution. Else, go to Step 2.

It can be seen that Step 2 computes a moment estimator of $\theta = (a, b)$ given (c, d), and Step 3 computes an unbiased estimator of (c, d) given θ. Step 2 demands the condition $(\bar{x} - c)(d - \bar{x}) > s$, and there is no proof of convergence of the iteration. In fact, in the following numerical studies by the authors, either of the condition or the convergence sometimes fails.

1.2.4 A New Estimator of the Generalized Beta Distributions

Nagatsuka, Balakrishnan, and Kamakura [21, 22] proposed the NagaBalaKama transform introduced in Sect. 1.2.1, but they used a conventional moment estimator for the location parameters. Hall and Wang [14] and Wang [27] proposed the use of empirical priors (1.2) and (1.6) for the estimation of the threshold, as we have seen in Sect. 1.2.2.

Our new estimator is based on a small modification of the empirical priors for the estimation of the location parameters. As seen, Hall and Wang's empirical prior

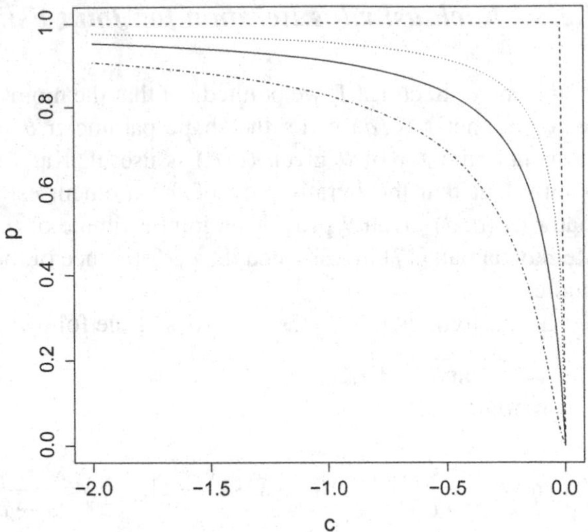

Fig. 1.1 Plots of $p = (p_1(c))^{\lambda_1}$ for different λ_1 with $x_{(1)} = 0$ and $x_{(2)} = 0.1$. Solid line: $\lambda_1 = 1$; broken line: $\lambda_1 \to 0$; dotted line: $\lambda_1 = 0.5$; chain: $\lambda_1 = 2$

(1.2) acts as a penalty depending on observations and the other parameters. We may further introduce a smoothing (or regularization) parameter $\lambda_1 > 0$ such that the penalty becomes $\lambda_1 \log(p_1(c))$, where if $\lambda_1 \to 0$, the estimation problem reduces to the original irregular case, while $\lambda_1 = 1$ gives the Hall and Wang's empirical prior. Note that $(p_1(c))^{\lambda_1}$ can still be regarded as a prior distribution. Figure 1.1 depicts $(p_1(c))^{\lambda_1}$ as a function of c for different λ_1.

Let $l(x; \theta, c, d, \lambda), \theta = (a, b), \lambda = (\lambda_1, \lambda_2)$, be our Bayesian log-likelihood, which is obtained from (1.8) by replacing the penalty terms $\log(p_1(c)) + \log(p_2(d))$ with $\lambda_1 \log(p_1(c)) + \lambda_2 \log(p_2(d))$, i.e.,

$$l(x; \theta, c, d, \lambda) = \lambda_1 \log(p_1(c)) + \lambda_2 \log(p_2(d)) + (\text{the other four terms}). \quad (1.9)$$

From the estimating equations for the location-scale parameter (c, d), $\partial l/\partial c = 0$ and $\partial l/\partial d = 0$, we obtain

$$a = 1 + \frac{\frac{nb}{d-c} - \lambda_1 \left(\frac{1}{x_{(1)}-c} - \frac{1}{x_{(2)}-c} \right)}{\sum_{i=1}^n \left(\frac{1}{x_{(i)}-c} - \frac{1}{d-c} \right)}, \quad b = 1 + \frac{\frac{na}{d-c} - \lambda_2 \left(\frac{1}{d-x_{(n)}} - \frac{1}{d-x_{(n-1)}} \right)}{\sum_{i=1}^n \left(\frac{1}{d-x_{(i)}} - \frac{1}{d-c} \right)}, \quad (1.10)$$

respectively. Note that, when $\lambda_1 = \lambda_2 = 0$, $a > 1$ and $b > 1$. Hence, as we have seen in Sect. 1.1, if the local maximum likelihood estimate exists, say $(\hat{\theta}, \hat{c}, \hat{d})$, then both of the estimates of the shape parameter $\hat{\theta} = (\hat{a}, \hat{b})$ must be greater than 1. The Eq. (1.10), being combined with $a > 0$ and $b > 0$, constrain the location-scale parameter:

$$\frac{d-c}{n}\sum_{i=1}^{n}\frac{1-\lambda_2(\delta_{i,n}-\delta_{i,n-1})}{d-x_{(i)}} > a+1, \quad \frac{d-c}{n}\sum_{i=1}^{n}\frac{1-\lambda_1(\delta_{i,1}-\delta_{i,2})}{x_{(i)}-c} > b+1.$$

In fact, it can be seen that if $\lambda_1 > 1$, then c cannot be close to $x_{(1)}$, and if $\lambda_2 > 1$, then d cannot be close to $x_{(n)}$ (see the chain in Fig. 1.1).

Our new estimator is obtained by the following two steps.

1. The shape parameter θ is estimated solely by the NagaBalaKama transform just once. The estimates $\hat{\theta}$ are the maximizers of the likelihood (1.7) free from the location-scale parameters (c, d).
2. The estimate of the location-scale parameter, say $(\hat{c}, \hat{d})$, is obtained by maximizing the Bayesian log-likelihood (1.9). The smoothing parameter $\hat{\lambda}$ can be determined by usual leave-one-out cross-validation, where we choose the minimizer of the empirical cross-entropy (or the Kolmogorov–Smirnov statistics, as in [17]) between the predictive distribution with estimated parameters and the true distribution (deviance of the predictive distribution):

$$(\hat{\lambda}_1, \hat{\lambda}_2) = \underset{\lambda_1,\lambda_2}{\operatorname{argmax}} \left\{ -\frac{1}{n}\sum_{i=1}^{n} \log f(x_i; \hat{\theta}, \hat{c}_{-i}(\lambda), \hat{d}_{-i}(\lambda)) \right\},$$

where $(\hat{c}_{-i}(\lambda), \hat{d}_{-i}(\lambda))$ is the maximum likelihood estimate obtained by the sample with the ith observation being removed.

1.3 Evaluation and Application to a Historical Tsunami Dataset

1.3.1 Evaluation of the New Estimator

The estimate of the shape parameter $\hat{\theta} = (\hat{a}, \hat{b})$ maximizes the likelihood (1.7). Although the authors do not have a rigorous proof of the existence, according to their experience of numerical studies, the existence seems to hold for a large n. Once the shape parameter is estimated, the estimate of the location-scale parameter, $(\hat{c}, \hat{d})$, is obtained by maximizing the Bayesian log-likelihood (1.9). In this subsection, we discuss the step of estimation of the location-scale parameter. For simplicity of notation, we omit hats from the shape parameter. In other words, we assume the shape parameter is known.

Existence

We have

$$\frac{\partial l}{\partial c} = -f(c) + \frac{n(a+b-1)}{d-c}, \quad \frac{\partial l}{\partial d} = g(d) - \frac{n(a+b-1)}{d-c},$$

and

$$\frac{\partial^2 l}{\partial c^2} = -f'(c) + \frac{n(a+b-1)}{(d-c)^2}, \quad \frac{\partial^2 l}{\partial d^2} = g'(d) + \frac{n(a+b-1)}{(d-c)^2}, \quad \frac{\partial^2 l}{\partial c \partial d} = -\frac{n(a+b-1)}{(d-c)^2},$$

where

$$f(c) = \sum_{i=1}^{n} \frac{a-1+\lambda_1(\delta_{i,1}-\delta_{i,2})}{x_{(i)}-c}, \tag{1.11}$$

$$g(d) = \sum_{i=1}^{n} \frac{b-1+\lambda_2(\delta_{i,n}-\delta_{i,n-1})}{d-x_{(i)}}. \tag{1.12}$$

The estimating equations $\partial l/\partial c = \partial l/\partial d = 0$ demand

$$f(\hat{c}) = g(\hat{d}) = \frac{n(a+b-1)}{\hat{d}-\hat{c}}, \quad \hat{c} < x_{(1)}, \quad \hat{d} > x_{(n)}. \tag{1.13}$$

There are so many possible cases for the joint region of parameters (a, b) and (λ_1, λ_2) which should be treated separately. There are two cases that the global maximum of the Bayesian log-likelihood (1.9) satisfying (1.13) exists: i) $a + b < 1$ and "$a > \max(0, 1 - \lambda_1)$ or $b > \max(0, 1 - \lambda_2)$"; ii) $\max(a, b) > 2$, $a > \max(0, 1 - \lambda_1)$, and $b > \max(0, 1 - \lambda_2)$. They are demonstrated as Propositions 1.3 and 1.4, respectively, in Appendix.

Remark The global maximum of the Bayesian log-likelihood (1.9) may satisfy neither of (1.13) nor $\hat{c} \in (-\infty, x_{(1)})$ and $\hat{d} \in (x_{(n)}, \infty)$. A well-known example is that for $a, b \leq 1$ and $\lambda_1 = \lambda_2 = 0$, we have $\hat{c} = x_{(1)}$ and $\hat{d} = x_{(n)}$.

Limit distributions

Let us consider the limit distribution of the estimator of the location-shape parameter $(\hat{c}, \hat{d})$ with known shape parameter $\theta = (a, b)$. If $\min(a, b) > 2$, the estimators of the location-scale parameters have the same asymptotic properties as in usual regular cases, since the smoothing terms in Bayesian log-likelihood (1.9) do not alter first-order asymptotic properties of the maximum likelihood estimator. Since the case of $\min(a, b) > 2$ without the smoothing terms was discussed thoroughly by Carnahan [6], we may concentrate on the cases with $\max(a, b) \leq 2$. In this paper, we discuss the cases of $(a, b) \in (0, 1)^2$ and $(a, b) \in (1, 2)^2$ by following Hall and Wang's arguments [14]. Other cases can be discussed similarly. The extremes $X_{(1)}$ and $X_{(n)}$ of an iid sample from the generalized beta distribution of pdf (1.5) are in the domain of attraction of the reverse Weibull distribution [10], whose distribution function is $\Psi_\alpha(x) = e^{-(-x)^\alpha}$, $x < 0$. In fact, it can be seen that

$$\eta_1 = \frac{(c - X_{(1)})n^{1/a}}{(B(a,b)a)^{1/a}(d-c)} \quad \text{and} \quad \zeta_1 = \frac{(X_{(n)} - d)n^{1/b}}{(B(a,b)b)^{1/b}(d-c)} \tag{1.14}$$

follow $\Psi_a(\cdot)$ and $\Psi_b(\cdot)$, respectively. The following theorem by Hall [12] is useful to discuss the limit distribution of the estimator of the location-scale parameter.

Theorem 1.1 ([12]) *Suppose an iid sample $(X_1, ..., X_n)$ from a distribution which is in the domain of attraction of the reverse Weibull distribution, whose distribution function is $\Psi_\alpha(x)$. Let the limit of the random vector*

$$((X_{(n)} - b_n)/a_n, ..., (X_{(n-r+1)} - b_n)/a_n)$$

be $(\xi_1, ..., \xi_r)$, $r \geq 1$, where a_n and b_n are sequences of constants chosen such that the vector has a non-degenerate limit. Define the random variables $\tilde{\xi}_i$ by

$$\tilde{\xi}_i^{(\alpha)} = -\exp\left[-\frac{1}{\alpha}\left\{\sum_{j=i}^{\infty}\frac{E_j - 1}{j} + \gamma - \sum_{j=1}^{i-1}\frac{1}{j}\right\}\right], \quad i \geq 1, \tag{1.15}$$

where the second summation is 0 if $i = 1$. Then, $(\xi_i; i \geq 1) \stackrel{d}{=} (\tilde{\xi}_i^{(\alpha)}; i \geq 1)$. Here, $E_1, E_2, ...$ are independent exponential random variables with mean 1, and γ is the Euler–Mascheroni constant.

Next, we prepare the following lemma on the asymptotic independence of the r-maxima and the r-minima of an iid sequence for each $r \geq 1$. Although the assertion is elementary, the authors could not find it in the literature. The case of $r = 1$ for an iid sequence was discussed in, for example, [24], and that for a stationary sequence was considered by Davis [8, 9] and a sufficient condition was given as Proposition 3.1 of [9].

Lemma 1.1 *For an iid sequence $(X_1, ..., X_n)$ following a continuous distribution, the random vectors*

$$((X_{(n)} - b_n)/a_n, ..., (X_{(n-r+1)} - b_n)/a_n), \quad ((X_{(1)} - b'_n)/a'_n, ..., (X_{(r)} - b'_n)/a'_n)$$

for each $r \geq 1$, where a_n, b_n, a'_n, and b'_n are sequences of constants chosen such that each vector has the non-degenerate limit, are asymptotically independent as $n \to \infty$.

Proof The independence of the two random vectors means

$$\begin{aligned}
&P(X_{(n)} \in I_1, ..., X_{(n-r+1)} \in I_r, X_{(1)} \in J_1, ..., X_{(r)} \in J_r) \\
&= P(X_{(n)} \in I_1, ..., X_{(n-r+1)} \in I_r)P(X_{(1)} \in J_1, ..., X_{(r)} \in J_r),
\end{aligned}$$

for all sets of intervals $I_i = (x_i^{(L)}, x_i^{(R)})$ with $x_{i+1}^{(R)} \leq x_i^{(L)}$ and $J_j = (y_j^{(L)}, y_j^{(R)})$ with $y_j^{(R)} \leq y_{j+1}^{(L)}$. We will consider the case that $x_{i+1}^{(R)} = x_i^{(L)} = x_i$, $i \in \{0, ..., r\}$, with $x_1^{(R)} = \infty$ and $x_r^{(L)} = x_r$, and $y_j^{(R)} = y_{j+1}^{(L)} = y_j$, $j \in \{0, ..., r\}$, with $y_1^{(L)} = -\infty$ and

$y_r^{(R)} = y_r$. Other cases can be shown in a similar manner. Let us introduce the counts of $(X_{(n)}, ..., X_{(n-r+1)})$ and $(X_{(1)}, ..., X_{(r)})$:

$$C_i = \#\{j : v_i = v_n(x_i) > X_j \geq v_n(x_{i+1}) = v_{i+1}, j \in \{1, ..., n\}\}$$
$$D_i = \#\{j : u_{i+1} = u_n(y_{i+1}) \geq X_j > u(y_i) = u_i, j \in \{1, ..., n\}\}, \quad i \in \{0, ..., r-1\},$$

where $u_n(x) = a_n x + b_n$ and $v_n(y) = a'_n y + b'_n$. To establish the asymptotic independence, it is sufficient to show that

$$P(C_0 = c_0, ..., C_{r-1} = c_{r-1}, D_0 = d_0, ..., D_{r-1} = d_{r-1})$$
$$\to P(C_0 = c_0, ..., C_{r-1} = c_{r-1})P(D_0 = d_0, ..., D_{r-1} = d_{r-1}), \quad n \to \infty. \tag{1.16}$$

The left-hand side of (1.16) leads to

$$\frac{[n]_{\sum_i (c_i+d_i)}}{\prod_{i=0}^{r-1} c_i! d_i!}(F(u_r) - F(v_r))^{n-\sum_i (c_i+d_i)} \prod_{i=0}^{r-1} (F(u_i) - F(u_{i+1}))^{d_i} (F(v_{i+1}) - F(v_i))^{c_i}$$
$$\to \frac{G(x_r)(1 - G'(y_r))}{\prod_{i=0}^{r-1} c_i! d_i!} \prod_{i=0}^{r-1} \log \frac{G(x_i)}{G(x_{i+1})} \log \frac{1 - G'(y_i)}{1 - G'(y_{i+1})},$$

where $[n]_i = n(n-1)\cdots(n-i+1)$ and $F(x) = P(X < x)$. To derive the limit, $n\{1 - F(u_n(x))\} \to -\log G(x)$ with $G(x) = P(X_{(n)} < x)$ and $-nF(v_n(y)) \to -\log(1 - G'(y))$ with $G'(y) = P(X_{(1)} < y)$ are used (see, for example, Sect. 1.1 of [11]). The second factor of the right-hand side of (1.16) leads to

$$\frac{[n]_{\sum_i d_i}}{\prod_{i=0}^{r-1} d_i!}(F(u_r))^{n-\sum_i d_i} \prod_{i=0}^{r-1} (F(u_i) - F(u_{i+1}))^{d_i} \to \frac{G(x_r)}{\prod_{i=0}^{r-1} d_i!} \prod_{i=0}^{r-1} \log \frac{G(x_i)}{G(x_{i+1})}.$$

The product of this limit and the limit of the first factor of the right-hand side of (1.16) equal the limit of the left-hand side of (1.16). □

The following proposition holds if $(a, b) \in (0, 1)^2$.

Proposition 1.1 *For the Bayesian log-likelihood (1.9) of an iid sample from the generalized beta distribution of the density (1.5) with a known shape parameter* $\theta = (a, b) \in (0, 1)^2$*, the maximum likelihood estimate of the location-scale parameter* $(\hat{c}, \hat{d})$ *satisfies*

$$\left(\frac{(c - \hat{c})n^{1/a}}{(B(a, b)a)^{1/a}(d - c)}, \frac{(\hat{d} - d)n^{1/b}}{(B(a, b)b)^{1/b}(d - c)} \right) \xrightarrow{d} (\eta, \zeta),$$

where (η, ζ) *is the solutions of the random equations* $\tilde{f}_1(\eta) = \tilde{g}_1(\zeta) = 0$*, which maximize (1.9) with*

$$\tilde{f}_1(\eta) = \sum_{i=1}^{\infty} \frac{a-1+\lambda_1(\delta_{i,1}-\delta_{i,2})}{\eta_i - \eta}, \quad \tilde{g}_1(\zeta) = \sum_{i=1}^{\infty} \frac{b-1+\lambda_2(\delta_{i,1}-\delta_{i,2})}{\zeta - \zeta_i}.$$

Here, the sequences of random coefficients $(\eta_i; i \geq 1)$ *and* $(\zeta_i; i \geq 1)$ *satisfy* $(\eta_i; i \geq 1) \stackrel{d}{=} (\tilde{\xi}_i^{(a)}; i \geq 1)$ *and* $(\zeta_i; i \geq 1) \stackrel{d}{=} (\tilde{\xi}_i^{(b)}; i \geq 1)$ *independently.*

Proof The estimating equations are asymptotically equivalent to

$$\sum_{i=1}^{n} \frac{(B(a,b)a)^{1/a}(d-c)\{a-1+\lambda_1(\delta_{i,1}-\delta_{i,2})\}}{\{(c-\hat{c})+(x_{(i)}-c)\}n^{1/a}} = 0$$

and

$$\sum_{i=1}^{n} \frac{(B(a,b)b)^{1/b}(d-c)\{b-1+\lambda_2(\delta_{n-i+1,n}-\delta_{n-i+1,n-1})\}}{\{(\hat{d}-d)+(d-x_{(n-i+1)})\}n^{1/b}} = 0$$

with $n \to \infty$. The solutions of truncated equations

$$\sum_{i=1}^{r} \frac{(B(a,b)a)^{1/a}(d-c)\{a-1+\lambda_1(\delta_{i,1}-\delta_{i,2})\}}{\{(c-\hat{c})+(x_{(i)}-c)\}n^{1/a}} = 0 \quad (1.17)$$

and

$$\sum_{i=1}^{r} \frac{(B(a,b)b)^{1/b}(d-c)\{b-1+\lambda_2(\delta_{n-i+1,n}-\delta_{n-i+1,n-1})\}}{\{(\hat{d}-d)+(d-x_{(n-i+1)})\}n^{1/a}} = 0 \quad (1.18)$$

are asymptotically independent for $r \geq 1$, since $(x_{(1)}, \ldots, x_{(r)})$ and $(x_{(n)}, \ldots, x_{(n-r+1)})$ are asymptotically independent by virtue of Lemma 1.1. Theorem 1.1 and (1.14) show that the distributions of the solutions of (1.17) and (1.18) converge to those of the solutions of the random equations $\tilde{f}_1^{(r)}(\eta) = 0$ and $\tilde{g}_1^{(r)}(\zeta) = 0$, respectively, where

$$\tilde{f}_1^{(r)}(\eta) = \sum_{i=1}^{r} \frac{a-1+\lambda_1(\delta_{i,1}-\delta_{i,2})}{\eta_i - \eta}, \quad \tilde{g}_1^{(r)}(\zeta) = \sum_{i=1}^{r} \frac{b-1+\lambda_2(\delta_{i,1}-\delta_{i,2})}{\zeta - \zeta_i},$$

$$\eta_i = \frac{(c-x_{(i)})n^{1/a}}{(B(a,b)a)^{1/a}(d-c)}, \quad \zeta_i = \frac{(x_{(n-i+1)}-d)n^{1/b}}{(B(a,b)b)^{1/b}(d-c)},$$

$$\eta = \frac{(c-\hat{c})n^{1/a}}{(B(a,b)a)^{1/a}(d-c)}, \quad \zeta = \frac{(\hat{d}-d)n^{1/b}}{(B(a,b)b)^{1/b}(d-c)},$$

and $(\eta_i; i \geq 1) \stackrel{d}{=} (\tilde{\xi}_i^{(a)}; i \geq 1)$ and $(\zeta_i; i \geq 1) \stackrel{d}{=} (\tilde{\xi}_i^{(b)}; i \geq 1)$. Finally, let $r \to \infty$.

For the case of $(a, b) \in (1, 2)^2$, a useful fact is that the random variable $1/X$, where X follows the generalized beta distribution of pdf (1.5), is in the domain of attraction of the stable distribution. In fact, it can be seen that

$$(B(a,b)a)^{1/a}\frac{d-c}{a-1}\left\{-\sum_{i=1}^{n}\frac{a-1}{(x_{(i)}-c)n^{1/a}}+\frac{a+b-1}{(d-c)n^{1/b-1}}\right\}\xrightarrow{d} Z_a \tag{1.19}$$

and

$$(B(a,b)b)^{1/b}\frac{d-c}{b-1}\left\{\sum_{i=1}^{n}\frac{b-1}{(d-x_{(n-i+1)})n^{1/b}}-\frac{a+b-1}{(d-c)n^{1/b-1}}\right\}\xrightarrow{d} Z_b, \tag{1.20}$$

where Z_α denotes the stable distribution of parameter α. The following proposition gives the asymptotic distribution of the estimator of the location-scale parameter. It can be proved in a similar way as Proposition 1.1, but the joint convergence of the sequence in Theorem 1.1, (1.19), and (1.20) is needed. See [5, 13, 14]. □

Proposition 1.2 *For the Bayesian log-likelihood (1.9) of an iid sample from the generalized beta distribution of the density (1.5) with a known shape parameter* $\theta = (a,b) \in (1,2)^2$, *the maximum likelihood estimate of the location-scale parameter* $(\hat{c},\hat{d})$ *satisfies*

$$\left(\frac{(c-\hat{c})n^{1/a}}{(B(a,b)a)^{1/a}(d-c)},\frac{(\hat{d}-d)n^{1/b}}{(B(a,b)b)^{1/b}(d-c)}\right)\xrightarrow{d}(\eta,\zeta),$$

where η *and* ζ *are the maximum likelihood solutions of the random equations* $\tilde{f}_2(\eta) = \tilde{g}_2(\zeta) = 0$ *which maximize (1.9) with*

$$\tilde{f}_2(\eta) = \frac{a-1+\lambda_2}{\eta_1-\eta}+\frac{a-1-\lambda_2}{\eta_2-\eta}+(a-1)\left(-\eta_1^{-1}-\eta_2^{-1}+\sum_{i=3}^{\infty}\frac{\eta}{\eta_i(\eta_i-\eta)}+Z_a\right).$$

$$\tilde{g}_2(\zeta) = \frac{b-1+\lambda_2}{\zeta-\zeta_1}+\frac{b-1-\lambda_2}{\zeta-\zeta_2}+(b-1)\left(\zeta_1^{-1}+\zeta_2^{-1}+\sum_{i=3}^{\infty}\frac{\zeta}{\zeta_i(\zeta-\zeta_i)}+Z_b\right).$$

Here, Z_a *and* Z_b *are independent random variables following the stable distribution of parameters a and b, respectively, and the sequences of random coefficients* $(\eta_i; i \geq 1)$ *and* $(\zeta_i; i \geq 1)$ *satisfy* $(\eta_i; i \geq 1) \stackrel{d}{=} (\tilde{\xi}_i^{(a)}; i \geq 1)$ *and* $(\zeta_i; i \geq 1) \stackrel{d}{=} (\tilde{\xi}_i^{(b)}; i \geq 1)$ *independently.*

Remark Based on a numerical study, Wang [27] anticipated that "we may expect that they ($\hat{c}$ and $\hat{d}$) are asymptotically independent". Propositions 1.1 and 1.2 confirm the anticipation. Note that if $\min(a,b) > 2$, $\hat{c}$ and $\hat{d}$ are dependent [6].

Numerical Evaluation

The estimates of parameters $(\hat{\theta},\hat{c},\hat{d})$ with the empirical priors (1.6) were computed for simulated data from the beta distribution with various shape parameter θ. The location-scale parameter was fixed as $(c,d) = (0,1)$ and the smoothing parameter was fixed as $(\lambda_1, \lambda_2) = (1, 1)$. To obtain the estimates of the shape parameter $\hat{\theta}$, we used the Nelder–Mead optimization, where the likelihood (1.7) was evaluated numerically in each iteration by the Gaussian quadrature implemented `integral2` in `pracma` package of `R` with some modification. The search was restricted to be $(\hat{a}, \hat{b}) \in (0, 100)^2$. To obtain the estimate of the location-scale parameter, $(\hat{c}, \hat{d})$, we used the L-BFGS-B algorithm implemented in `optim` of `R`, where the search was restricted to be $\hat{c} \in (-5, x_{(1)})$ and $\hat{d} \in (x_{(n)}, 6)$. Initial values were set to be the true values. The bias and root mean squared errors (RMSE) of the estimators are tabulated in Tables 1.1 ($n = 30$) and 1.2 ($n = 100$). We also implemented Wang's method introduced in Sect. 1.2.2 and Kachiashvili et al.'s method introduced in Sect. 1.2.3. The new method and Kachiashvili et al.'s method are indicated as NBK and Kach, respectively. As previously mentioned, Wang's method frequently fails to yield the estimates for small n (see Sect. 1.2.2) and Kachiashvili et al.'s method also some times fails (see Sect. 1.2.3). Table 1.3 shows the frequency with which the estimates were not available. Due to the poor property of Wang's method, comparison of the performance of the methods including Wang's method was quite fragile. Therefore, for each simulated data, we firstly applied Wang's method, and if Wang's method successfully found the estimates, then we applied Kachiashvili et al.'s method. If Kachiashvili et al.'s method also founds the estimates, the new method was applied. The new method also failed in some instances, but the frequency was significantly smaller than that of Wang's method and Kachiashvili et al.'s method. The number of replicates, 1000, in Tables 1.1 and 1.2 is the number of simulated data for which all of the three methods successfully gave the estimates. From these results, we note the following.

- For the estimation of the location-scale parameter, the new estimator almost always overperformed Wang's estimator in terms of both the bias and RMSE. For the shape parameter, the new estimator overperformed Wang's estimator if either a or b was large (larger than or equal to 2). The latter trend on the shape parameter is consistent with the observations by Nagatsuka et al. in the three-parameter gamma and Weibull distributions [21, 22].
- For $n = 30$ if either a or b was large (larger than or equal with 2), the RMSE of Wang's estimator and that of Kachiashvili et al.'s estimator were huge. This observation is consistent with the finding by Carnahan [6] for $\min(a, b) > 2$ (the limit distribution of Wang's estimator is identical to that of Carnahan's estimator). Carnahan concluded that "only for large samples ($n \geq 1000$), the bias in the estimates becomes small and the Cramér–Rao bound gives a good approximation for their variance".
- Kachiashvili et al.'s estimator is significantly biased if either a or b was large (equal with 5), as was reported in Fig. 1 and Table 2 of [17].

Note that the results displayed in Tables 1.1 and 1.2 are conditional; they are based on simulated data for which all of Wang's method, Kachiashvili et al.'s, and the new method successfully provided the estimates. In addition, the actual performance of Wang's estimator and the new estimator could be worse if local maxima appear, because their iterations started from the true value. As Table 1.3 shows, a significant

Table 1.1 Bias and RMSE of estimates with $c = 0, d = 1, n = 30$. Number of replications: 1000

		$\hat{a}$		$\hat{b}$		$\hat{c}$		$\hat{d}$	
(a, b)	Method	Bias	RMSE	Bias	RMSE	Bias	RMSE	Bias	RMSE
(0.5, 0.5)	Kach	0.008	0.176	0.005	0.201	-0.003	0.015	0.003	0.021
	Wang	0.048	0.170	0.044	0.175	0.001	0.012	-0.001	0.014
	NBK	0.073	0.199	0.069	0.220	0.000	0.013	-0.000	0.019
(1, 1)	Kach	0.162	1.105	0.187	1.664	-0.002	0.084	0.020	0.132
	Wang	0.188	1.647	0.235	2.360	-0.005	0.094	0.011	0.213
	NBK	0.251	0.895	0.288	2.426	-0.010	0.077	0.011	0.177
(2, 2)	Kach	1.505	7.433	1.592	8.198	-0.074	0.334	0.076	0.346
	Wang	1.713	6.743	1.938	7.835	-0.077	0.389	0.089	0.445
	NBK	0.754	4.533	0.681	3.689	-0.029	0.277	0.023	0.232
(5, 5)	Kach	1.853	11.281	1.630	11.347	-0.051	0.421	0.031	0.415
	Wang	4.975	16.440	4.638	16.016	-0.140	0.644	0.118	0.603
	NBK	0.050	2.119	-0.100	0.887	-0.003	0.183	-0.013	0.110
(1, 0.5)	Kach	0.194	2.706	0.020	0.239	-0.040	0.336	0.002	0.009
	Wang	0.165	0.818	0.044	0.180	-0.009	0.144	-0.000	0.006
	NBK	0.380	2.033	0.066	0.198	-0.044	0.333	-0.000	0.006
(2, 0.5)	Kach	0.932	5.759	0.017	0.195	-0.222	1.294	0.001	0.004
	Wang	0.739	3.030	0.033	0.142	-0.141	0.720	-0.000	0.002
	NBK	0.472	2.061	0.038	0.148	-0.072	0.484	-0.000	0.002
(5, 0.5)	Kach	-0.267	10.070	-0.038	0.164	0.093	1.301	0.000	0.001
	Wang	0.927	8.315	0.010	0.119	-0.035	1.090	-0.000	0.001
	NBK	0.062	1.216	0.033	0.125	0.032	0.277	-0.000	0.001
(2, 1)	Kach	0.961	5.593	0.108	0.772	-0.120	0.685	0.007	0.038
	Wang	1.303	5.779	0.117	0.763	-0.139	0.683	-0.000	0.034
	NBK	0.959	4.267	0.138	0.563	-0.093	0.505	-0.000	0.027
(5, 1)	Kach	0.852	10.804	-0.041	0.625	-0.058	1.276	0.002	0.015
	Wang	1.566	9.859	-0.001	0.866	-0.096	1.035	-0.001	0.019
	NBK	0.159	2.987	0.090	0.641	0.024	0.293	-0.001	0.014
(5, 2)	Kach	2.125	12.041	0.161	1.903	-0.130	1.006	0.007	0.073
	Wang	5.203	17.713	0.570	3.352	-0.254	1.068	0.006	0.094
	NBK	0.095	2.570	0.225	1.453	0.025	0.189	0.000	0.056

proportion of simulated data were discarded, especially if either a or b is large (larger than or equal to 2). Therefore, the actual performance of these estimators should be worse than the results displayed in Tables 1.1 and 1.2. In fact, the performance observed in Tables 1.4 and 1.5 for the cases of $(\lambda_1, \lambda_2) = (1, 1)$ are worse than that in Tables 1.1 and 1.2.

To see the influence of the choice of the smoothing parameter λ on the performance of the new estimator, the estimates of parameters $(\hat{\theta}, \hat{c}, \hat{d})$ with the Bayesian log-likelihood (1.9) were computed for simulated data from the beta distribution with

Table 1.2 Bias and RMSE of estimates with $c = 0, d = 1, n = 100$. Number of replications: 1000

		$\hat{a}$		$\hat{b}$		$\hat{c}$		$\hat{d}$	
(a, b)	Method	Bias	RMSE	Bias	RMSE	Bias	RMSE	Bias	RMSE
(0.5, 0.5)	Kach	0.001	0.084	0.001	0.084	-0.000	0.001	0.000	0.001
	Wang	0.016	0.069	0.016	0.069	0.000	0.001	-0.000	0.001
	NBK	0.016	0.074	0.017	0.074	0.000	0.001	-0.000	0.001
(1, 1)	Kach	0.007	0.182	0.004	0.184	-0.001	0.012	0.001	0.012
	Wang	-0.004	0.161	-0.008	0.167	0.004	0.011	-0.004	0.012
	NBK	0.045	0.181	0.040	0.184	0.002	0.012	-0.002	0.012
(2, 2)	Kach	0.006	0.659	-0.011	0.641	-0.004	0.055	0.001	0.053
	Wang	0.002	0.695	-0.018	0.652	0.005	0.058	-0.008	0.053
	NBK	0.045	0.631	0.025	0.585	0.003	0.055	-0.005	0.049
(5, 5)	Kach	1.358	8.757	1.354	9.552	-0.033	0.274	0.031	0.287
	Wang	0.829	5.034	0.780	5.049	-0.014	0.205	0.011	0.199
	NBK	0.130	3.507	0.104	4.210	-0.002	0.151	-0.002	0.173
(1, 0.5)	Kach	0.001	0.185	0.004	0.084	-0.002	0.024	0.000	0.001
	Wang	0.019	0.175	0.016	0.066	0.006	0.024	-0.000	0.000
	NBK	0.049	0.194	0.018	0.072	0.004	0.025	-0.000	0.000
(2, 0.5)	Kach	-0.005	0.826	-0.009	0.089	-0.010	0.173	0.000	0.000
	Wang	0.191	1.078	0.016	0.065	-0.020	0.252	0.000	0.000
	NBK	0.214	1.180	0.009	0.068	-0.036	0.284	-0.000	0.000
(5, 0.5)	Kach	-0.132	6.214	-0.022	0.094	-0.004	0.965	-0.000	0.000
	Wang	1.957	6.721	0.026	0.065	-0.247	1.001	0.000	0.000
	NBK	0.186	2.428	0.011	0.058	-0.014	0.297	-0.000	0.000
(2, 1)	Kach	0.050	0.713	0.010	0.200	-0.009	0.102	0.001	0.007
	Wang	0.008	0.685	-0.006	0.173	0.009	0.104	-0.002	0.006
	NBK	0.118	0.753	0.031	0.180	-0.003	0.114	-0.002	0.006
(5, 1)	Kach	2.014	11.312	-0.032	0.221	-0.256	1.406	0.000	0.003
	Wang	0.681	5.201	-0.029	0.166	-0.076	0.651	-0.001	0.002
	NBK	0.506	5.025	0.008	0.144	-0.042	0.516	-0.001	0.002
(5, 2)	Kach	2.253	11.004	0.028	0.753	-0.184	0.908	0.002	0.025
	Wang	0.843	6.044	-0.053	0.644	-0.059	0.500	-0.005	0.023
	NBK	0.271	3.850	-0.014	0.537	-0.014	0.312	-0.003	0.022

various shape parameter θ and smoothing parameter λ. The bias and RMSE of the estimators are tabulated in Tables 1.4 ($(a, b) = (5, 5), (2, 2), (1, 1), (0.5, 0.5)$) and Table 1.5 ($(a, b) = (2, 1)$).

In the last column, the bias is the sum of absolute values of the biases of $(\hat{c}, \hat{d})$, and the RMSE is the square root of the sum of MSEs of $(\hat{c}, \hat{d})$. From these results, we note the following.

Table 1.3 Frequency with which the parameter estimates were not available in producing Tables 1.1 and 1.2

n	30			100		
(a, b)	Wang	Kach	NBK	Wang	Kach	NBK
(0.5, 0.5)	9	0	0	2	0	0
(1, 1)	4	0	9	2	0	0
(2, 2)	13	9	15	1	1	0
(5, 5)	91	43	20	49	22	8
(1, 0.5)	6	1	0	1	0	0
(2, 0.5)	81	5	3	0	0	0
(5, 0.5)	337	14	0	163	14	1
(2, 1)	31	9	11	2	0	0
(5, 1)	213	66	31	47	26	8
(5, 2)	111	53	28	10	54	6

Table 1.4 Bias and RMSE of estimates with $c = 0, d = 1, n = 100$. Number of replications: 1000

		$\hat{a}$		$\hat{b}$		$\hat{c}$		$\hat{d}$		Loc.-Scale	
(a, b)	(λ_1, λ_2)	Bias	RMSE	Bias	RMSE	Bias	RMSE	Bias	RMSE	Bias	RMSE
(5, 5)	(2, 2)	0.380	6.002	0.307	4.977	-0.014	0.197	-0.007	0.237	0.022	0.308
	(1, 1)					-0.012	0.197	-0.009	0.236	0.022	0.307
	(0.5, 0.5)					-0.010	0.196	-0.012	0.236	0.022	0.307
	(0, 0)					-0.008	0.196	-0.014	0.235	0.022	0.306
	Cramér-Rao		3.033		3.033		0.143		0.143		0.202
(2, 2)	(2, 2)	0.040	0.642	0.042	0.624	-0.009	0.055	0.007	0.063	0.016	0.084
	(1, 1)					-0.005	0.055	0.004	0.063	0.009	0.083
	(.5, .5)					-0.001	0.054	-0.000	0.062	0.001	0.082
	(0, 0)					0.004	0.054	-0.005	0.062	0.009	0.082
(1, 1)	(2, 2)	0.046	0.178	0.044	0.177	-0.007	0.015	0.005	0.015	0.012	0.022
	(1, 1)					-0.004	0.014	0.003	0.014	0.007	0.019
	(0.5, 0.5)					-0.001	0.012	0.000	0.013	0.002	0.018
	(0, 0)					0.002	0.011	-0.003	0.012	0.005	0.017
(0.5, 0.5)	(2, 2)	0.017	0.076	0.013	0.070	-0.001	0.002	0.001	0.002	0.001	0.002
	(1, 1)					-0.000	0.001	-0.000	0.001	0.001	0.002
	(0.5, 0.5)					-0.000	0.001	0.000	0.001	0.000	0.002
	(0, 0)					0.000	0.001	-0.000	0.001	0.000	0.002

Table 1.5 Bias and RMSE of estimates with $(a, b, c, d) = (2, 1, 0, 1), n = 100$. Number of replications: 1000. Bias and RMSE of $\hat{a}$ is 0.114 and 0.693, respectively, and Bias and RMSE of $\hat{b}$ is 0.030 and 0.186, respectively

	$\hat{c}$		$\hat{d}$		Loc.-Scale	
(λ_1, λ_2)	Bias	RMSE	Bias	RMSE	Bias	RMSE
(2, 2)	-0.022	0.107	0.002	0.033	0.024	0.111
(2, 1)	-0.022	0.106	0.001	0.032	0.022	0.111
(2, 0.5)	-0.021	0.105	-0.001	0.032	0.022	0.110
(2, 0)	-0.020	0.105	-0.002	0.032	0.022	0.110
(1, 2)	-0.017	0.106	0.002	0.033	0.019	0.110
(1, 1)	-0.017	0.105	0.001	0.032	0.017	0.110
(1, 0.5)	-0.016	0.104	-0.001	0.032	0.017	0.109
(1, 0)	-0.015	0.104	-0.002	0.032	0.017	0.109
(0.5, 2)	-0.012	0.105	0.002	0.033	0.014	0.110
(0.5, 1)	-0.011	0.104	0.001	0.032	0.012	0.109
(0.5, 0.5)	-0.010	0.104	-0.001	0.032	0.011	0.109
(0.5, 0)	-0.010	0.103	-0.002	0.032	0.012	0.108
(0, 2)	-0.005	0.105	0.002	0.033	0.007	0.110
(0, 1)	-0.005	0.104	0.001	0.032	0.005	0.109
(0, 0.5)	-0.004	0.104	-0.001	0.032	0.004	0.109
(0, 0)	-0.003	0.103	-0.002	0.032	0.005	0.108

- When $a = b$, the RMSE of the location-scale parameter increases with increase of λ, while the bias once decreases and then increases. Without smoothing ($\lambda = (0, 0)$), $\hat{c}$ has positive bias and $\hat{d}$ has negative bias when $a = b \leq 2$. This trend is anticipated since $\hat{c} = x_{(1)}$ and $\hat{d} = x_{(n)}$ if $a = b \leq 1$.
- When $(a, b) = (2, 1)$, the RMSE of the location-scale parameter increases with increase in λ. For each fixed λ_1 (λ_2), the bias of location-scale parameter once decreases and then increases with increase in λ_2 (λ_1). The bias of $\hat{d}$ is negative if λ_2 is small and positive if λ_2 is large, while that of $\hat{c}$ is always negative.

The behavior of bias of the location-scale parameter is anticipated from the property of the smoothing terms observed in Fig. 1.1 (see Sect. 1.2.4), but we did not expect RMSEs to increase with the increase of the smoothing parameters λ. The empirical prior (1.6) resolves the bias coming from the non-regularity of the estimation, but it would not be optimal from the viewpoint of prediction. This finding strongly motivates us to choose the optimal smoothing parameter λ.

1.3.2 Fitting the Generalized Beta Distribution to a Historical Tsunami Dataset

Seismic magnitude scales measure the strength of earthquakes, and several scales are used according to the purpose of the scale used and the method of measurement. The Richter magnitude scale was once dominant; it is based on the maximum amplitude of horizontal ground-shaking. Recent magnitude scales tend to measure the energy of earthquakes, however, their values are still close to the Richter scale to match the shaking.

Abe considered measuring the scale of earthquakes that caused a large tsunami, and proposed *the tsunami magnitude scale* M_t, based on the tsunami-wave amplitude caused by an ocean bed earthquake, measured by tide gauges, [1]. His catalog of tsunami magnitudes, [2], contained 160 tsunamis observed in Japan from 1894 to 2006. They were all available data since the start of daily tide level measurement at regular times. The dataset was analyzed by Sibuya and Takahashi [25] fitting the generalized Pareto distribution, based on the extreme value statistics theory.

Later, Abe [3] remarked that run-up height, i.e., the height reached by a tsunami on the ground above sea level, can replace the tsunami-wave amplitude, and he extended his catalog to an earlier time, surveying documents, monuments, ruins, and geological layers. The new catalog contained 21 records and went back to 1498.

In reality, the data from the earlier period of his"modern" catalog (1894–2006) are incomplete; the number of tide gauges is less, and missing data are not negligible. Hence, the "historical" period includes 1498–1920, and the number of records is augmented to 31 with the range M_t, (6.7, 8.6). The year-M_t pairs are plotted in Fig. 1.2, which suggests that records of smaller magnitudes disappear faster from people's memory.

An advanced approach to natural disasters assumes time-position pair of an event, $\{(t, x)\}$, as the Poisson point process, homogeneous or nonhomogeneous, and the loss or size of the event Y, as the marked point process $\{Y(t, x)\}$. See Ogata (2015).

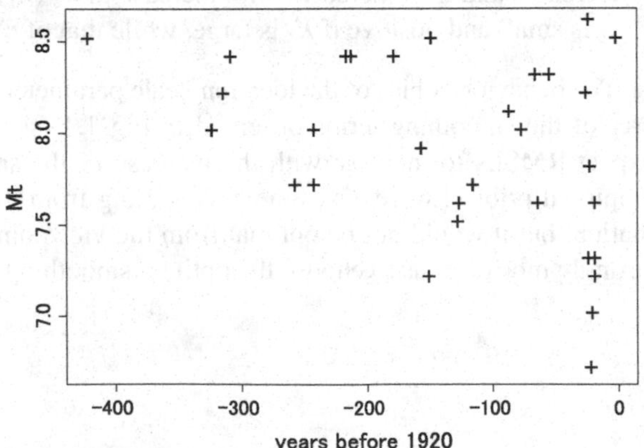

Fig. 1.2 Historical Tsunami-Magnitude dataset: M_t versus years before 1920

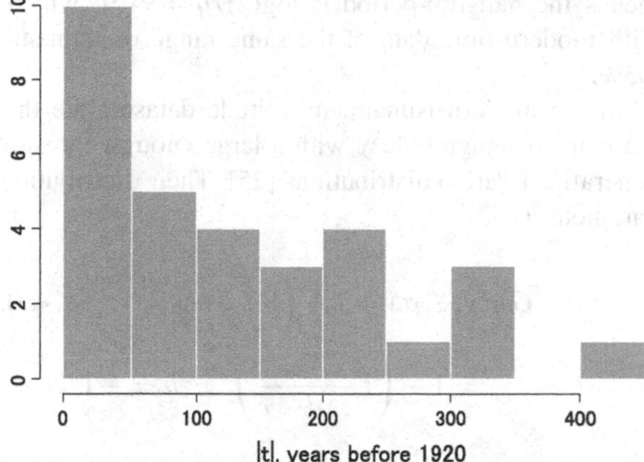

Fig. 1.3 Years $|t|$ of the historical dataset, t: years before 1920

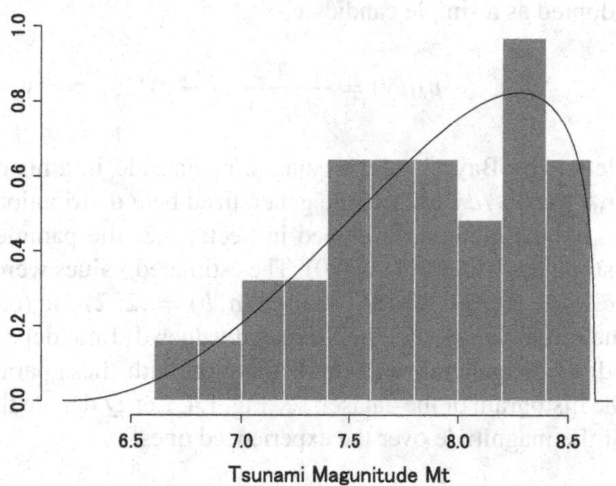

Fig. 1.4 Histogram of tsunami magnitude, $y = M_t$, of historical dataset, and the fitted generalized beta density

Here, for simplicity, we assume the year and tsunami magnitude pair, $(t, y = M_t)$, is a Poisson point process, which is t-homogeneous, and that t and y are independent. The availability w_i of the event (t_i, y_i) at time $t = 0$ is TRUE(= 1) or FALSE(= 0), and we assume the probability $P\{w_i = \text{TRUE}|t_i, y_i\} =: p(t_i, y_i)$ to decrease to 0 when $|t_i| \to \infty$ and increase to 1 when $y_i \to \infty$. That is, the historical dataset is the conditional process $(t, y|w = 1)$ of the marked Poisson point process. Figure 1.3 suggests the exponentiality of $|t|$, and in fact all q-values of the exponentiality tests in the R package are not small (it appears that tsunamis with smaller magnitude are intentionally or unintentionally neglected). Hence, the question is the marginal distribution of $y = M_t$. See, its histogram, Fig. 1.4.

Further, we assume the persisting probability is factorized $p(t_i, y_i) = p_t(t_i)p_M(y_i)$. Then, the marginal distribution $P\{w_i = 1|y_i\} = p_M(y_i)$ can be analyzed disregarding the other marginal $P\{w_i = 1|t_i\} = p_t(t_i)$. The missing data is time-homogeneous Poisson process, with intensity per year, say, η, and $1/\hat{\eta} = |\bar{t}| = 132.6$, which means the half-life-period is $\log(2)/\hat{\eta} = 91.9$, which looks plausible. Compared with modern-time data of the same range of tsunami magnitude, the intensity is 5.3%.

In the modern tsunami magnitude dataset, the threshold excess $x = y - c$ of the tsunami magnitude y with a large enough threshold c can be modeled by the generalized Pareto distributions [25]. Their distribution function with the restricted parameter is

$$G_P(y; \xi, \sigma) = 1 - \left(1 + \xi \frac{x}{\sigma}\right)^{-1/\xi}, \quad \xi < 0,\ 0 < x < -\sigma/\xi,$$
$$= 1 - \left(1 - \frac{y - c}{d - c}\right)^{b}, \quad b = -1/\xi,\ d = c - \sigma/\xi.$$

It is reasonable to assume the range of $p_M(y)$ equal to (c, d), and the following is adopted as a simple candidate:

$$p_M(y) = \frac{a}{(d - c)^a}(y - c)^{a-1}, \ c < y < d, \ a > 0.$$

Hence, by Bayes' rule, a tsunami magnitude in a historical record $f(y|w = 1) \propto p_M(y) dG_P/dy$ follows the generalized beta distribution (1.5).

By the method developed in Sect. 1.3.2, the parameters (a, b) and (c, d) were estimated, with λ=(0.9, 0.0). The estimated values were $(\hat{a}, \hat{b})$=(2.885, 1.317) and $(\hat{c}, \hat{d}) = (6.253, 8.618)$. We used $(a, b) = (2, 2)$ and $(c, d) = (x_{(1)} - 1, x_{(n)} + 1)$ as the initial values, but the estimated values did not depend on the initial values. The pdf of the generalized beta distribution with these parameters is shown along with the histogram of the dataset; see Fig. 1.4. Let Q denote the probability of the tsunami of the magnitude over the experienced ones

$$Q := P\{Y > y_{(n)}; a, b, c, d\} = P\left\{X > \frac{y_{(n)} - c}{d - c}; a, b\right\}.$$

The plug-in estimate of Q, at 1920, over the record $Mt = 8.6$ was $\hat{Q} = 0.0057$, a small but non-negligible value.

Let us see the confidence interval of $(\hat{c}, \hat{d})$. In Sect. 1.3.1, we have discussed the asymptotic distribution of the estimator of the location-scale parameter $(\hat{c}, \hat{d})$ when the shape parameter $\theta = (a, b)$ are known. Here, we are interested in the confidence interval when θ are unknown. Moreover, the asymptotic distributions displayed there are too complicated for actual use. A practical alternative is the use of the bootstrap. Figure 1.5 shows $(\hat{c}, \hat{d})$ obtained by the parametric bootstrap with 500 resamplings. The double bootstrap has an advantage in terms of asymptotic convergence rate [20]. The double bootstrap estimates of the marginal confidence intervals of $\hat{c}$ and $\hat{d}$ were $[-2.912, 7.201]$ and $[8.388, 8.842]$, respectively, where 200 bootstrap resamplings were conducted for each of the samples obtained by the 500 bootstrap resamplings

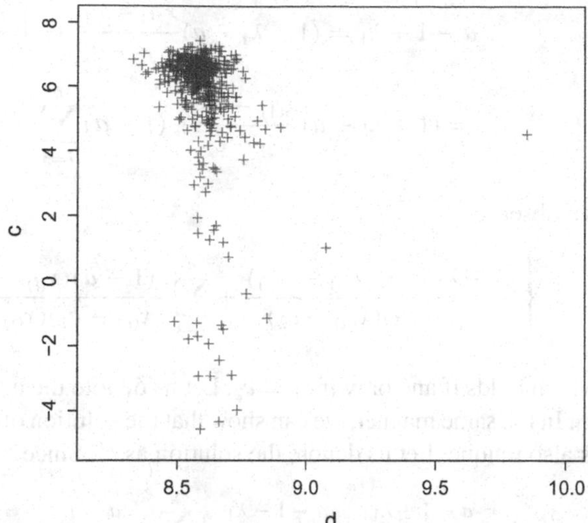

Fig. 1.5 Plot of $(\hat{c}, \hat{d})$ obtained by 500 bootstrap resamplings

to calibrate the confidence levels. The long confidence interval of $\hat{c}$ may come from a property of historical data, where the threshold to be recorded is generally ambiguous.

Acknowledgements The authors would like to thank the referee for drawing their attention to Kachiashvili et al. works [16, 17].

Appendix

Proposition 1.3 *For the Bayesian log-likelihood (1.9) of an iid sample* $(x_1, ..., x_n)$ *from the generalized beta distribution of the density (1.5) with shape parameter* $\theta = (a, b)$, $a + b < 1$, *and "*$a > \max(0, 1 - \lambda_1)$ *or* $b > \max(0, 1 - \lambda_2)$*" known, there exists the red global maximum likelihood estimate of the location-scale parameter* $(\hat{c}, \hat{d})$ *such that* $\hat{c} \in (-\infty, x_{(1)})$ *and* $\hat{d} \in (x_{(n)}, \infty)$ *for large n.*

Proof Let us consider the continuous function (1.11), that is,

$$f(c) = \frac{a - 1 + \lambda_1}{x_{(1)} - c} + \frac{a - 1 - \lambda_1}{x_{(2)} - c} + \sum_{i=3}^{n} \frac{a - 1}{x_{(i)} - c}.$$

Since $f(c) \sim (a - 1 + \lambda_1)/(x_{(1)} - c) \to +\infty$ as $c \uparrow x_{(1)}$ and $f(c) \sim n(1 - a)/c \to -0$ as $c \to -\infty$, there exists a solution of $f(c) = 0$ in $c \in (-\infty, x_{(1)})$. The solution is unique as follows. Suppose we have two solutions of $f(c) = 0$, say c_1 and c_2. Since

$$a-1+\lambda_1=(1+\lambda_1-a)\frac{x_{(1)}-c_1}{x_{(2)}-c_1}+(1-a)\sum_{i=3}^{n}\frac{x_{(1)}-c_1}{x_{(i)}-c_1}$$
$$=(1+\lambda_1-a)\frac{x_{(1)}-c_2}{x_{(2)}-c_2}+(1-a)\sum_{i=3}^{n}\frac{x_{(1)}-c_2}{x_{(i)}-c_2},$$

we observe

$$\left\{\frac{(1+\lambda_1-a)(x_{(2)}-x_{(1)})}{(x_{(2)}-c_1)(x_{(2)}-c_2)}+\sum_{i=3}^{n}\frac{(1-a)(x_{(i)}-x_{(1)})}{(x_{(i)}-c_1)(x_{(i)}-c_2)}\right\}(c_2-c_1)=0,$$

which holds if and only if $c_1=c_2$. Let us denote the unique solution of $f(c)=0$ as c_0. In the same manner, we can show that the solution of $f'(c)=0$ in $c\in(-\infty,x_{(1)})$ is also unique. Let us denote the solution as c'. Since

$$f'(c_0)=\frac{a-1+\lambda_1}{(x_{(1)}-c_0)^2}+\frac{a-1-\lambda_1}{(x_{(2)}-c_0)^2}+\sum_{i=3}^{n}\frac{a-1}{(x_{(i)}-c_0)^2}$$
$$>\frac{1-a}{a-1+\lambda_1}\left\{\frac{2(1+\lambda_1-a)}{(x_{(2)}-c_0)^2}+\sum_{i=3}^{n}\frac{1}{x_{(i)}-c_0}\left(\frac{2-\lambda_1-2a}{x_{(i)}-c_0}+\frac{2(1+\lambda_1-a)}{x_{(2)}-c_0}\right)\right\}$$
$$>\frac{1-a}{a-1+\lambda_1}\left\{\frac{2(1+\lambda_1-a)}{(x_{(2)}-c_0)^2}+\sum_{i=3}^{n}\frac{1}{x_{(i)}-c_0}\left(\frac{-\lambda_1}{x_{(i)}-c_0}+\frac{2(1+\lambda_1-a)}{x_{(2)}-c_0}\right)\right\}$$
$$>\frac{1-a}{a-1+\lambda_1}\left\{\frac{2(1+\lambda_1-a)}{(x_{(2)}-c_0)^2}+\sum_{i=3}^{n}\frac{2+\lambda_1-2a}{(x_{(i)}-c_0)(x_{(2)}-c_0)}\right\}>0,$$

we have $c'\in(-\infty,c_0)$ with $f(c')<0$. In the same manner, it can be seen that $g(d)=0$ has the unique solution $d_0\in(x_{(n)},\infty)$, and $g'(d)=0$ has the unique solution $d'\in(d_0,\infty)$ with $g(d')<0$. By using the properties of the functions $f(c)$ and $g(d)$ we have seen so far, we can choose $(\tilde{c},\tilde{d})$ such that

$$f(\tilde{c})=g(\tilde{d})=\frac{n(a+b-1)}{\hat{d}-\hat{c}}<0,\quad f'(\tilde{c}),g'(\tilde{d})>0$$

with $\tilde{c}\in(c',c_0)$ and $\tilde{d}\in(d',\infty)$. The Hessian is negative definite at $(\tilde{c},\tilde{d})$, because the determinant is negative:

$$-f'(\tilde{c})g'(\tilde{d})+\frac{n(a+b-1)}{(\tilde{d}-\tilde{c})^2}\{g'(\tilde{d})-f'(\tilde{c})\}<0$$

for large n. Here, the first term dominates since $f'(\tilde{c})=O(n^{2/a})$ and $g'(\tilde{d})=O(n^{2/b})$ (see Theorem 1.1). Hence, $(\tilde{c},\tilde{d})$ is a maximum of the Bayesian log-likelihood (1.9) for large n.

Likewise, we have another maximum $(\tilde{c}', \tilde{d}')$ with $f'(\tilde{c}'), g'(\tilde{d}') < 0$, and either of $(\tilde{c}, \tilde{d})$ or $(\tilde{c}', \tilde{d}')$ is the global maximum likelihood estimate of the location-scale parameter. □

Proposition 1.4 *For the Bayesian log-likelihood (1.9) of an iid sample* $(x_1, \ldots, x_n)$ *from the generalized beta distribution of the density (1.5) with shape parameter* $\theta = (a, b)$, $\max(a, b) > 2$, $a > \max(0, 1 - \lambda_1)$, *and* $b > \max(0, 1 - \lambda_2)$ *known, there exists the global maximum likelihood estimate of the location-scale parameter* $(\hat{c}, \hat{d})$ *such that* $\hat{c} \in (-\infty, x_{(1)})$ *and* $\hat{d} \in (x_{(n)}, \infty)$ *for large n.*

The proof is omitted since this proposition can be shown in a similar manner to the proof of Proposition 1.3.

References

1. Abe K (1979) Size of great earthquakes of 1837–1974 inferred from tsunami data. J Geophys Res 84:1561–1568
2. Abe K (1981) Physical size of tsunamigenic earthquakes of the northwestern Pacific. Phys. Earth Planet Inter. 27:194–205
3. Abe K (1999) Quantification of historical tsunamis by the M_t scale. Zisin, 2nd Series 52: 369–377 (in Japanese with English abstract)
4. Akahira M (1975) Asymptotic theory for estimation of location in non-regular case, I: Order of convergence of consistent estimators. Rep Statis. Appl Res, Union Jap Sci Eng 22:8–26
5. Anderson CW, Turkman KF (1991) The joint limiting distribution of sums and maxima of stationary sequences. J Appl Probab 28:33–44
6. Carnahan JV (1989) Maximum likelihood estimation for the 4-parameter beta distribution. Commun Statist Simul Comp 18:513–536
7. Cheng RCH, Iles TC (1987) Corrected maximum likelihood in nonregular estimation. J Roy Statist Soc B 49:95–101
8. Davis RA (1979) Maxima and minima of stationary sequences. Ann Probab 7:453–460
9. Davis RA (1982) Limit laws for the maximum and minimum of stationary sequences. Z Wahrschenlichkeitstheorie verw Gebiete 61:31–42
10. Gnedenko BV, (1943) Sur la distribution limite du terme maximum d'une série alétoire. Ann Math 44: 423–453; English translation: Kotz S, Johnson NL, (eds) (1991) Breakthroughs in Statistics. Foundations and Basic Theory. Springer, New York, pp 195–225
11. de Haan L, Ferreira A (2006) Extreme Value Theory: An Introduction. Springer, New York
12. Hall P (1978) Representations and limit theorems for extreme value distributions. J Appl Probab 15:639–644
13. Hall P (1978) On the extreme terms of a sample from the domain of attraction of a stable law. J London Math Soc 18:181–191
14. Hall P, Wang JZ (2005) Bayesian likelihood methods for estimation the end point of a distribution. J Roy Satist Soc B 67:717–729
15. Johnson NL, Kotz S, Balakrishnan N (1995) Continuous univariate distributions, vol 2, 2nd edn. Wiley, New York, chap 25
16. Kachiashvili KJ, Prangishvili AI (2018) Verification in biometric systems: problems and modern methods of their solution. J Appl Stat 45:43–62
17. Kachiashvili KJ, Melikdzhanjan DI (2019) Estimators of the parameters of Beta distributions. Shankhyā 81B:350–373
18. Le Cam L (1969) Theorie asymptotique de la decision statistique. Les Presses de L'Université Montreal

19. Le Cam L (1970) On the assumptions used to prove asymptotic normality of maximum likelihood estimates. Ann Statist 41:802–828
20. Martin MA (1992) On the Double Bootstrap. In: Page C, LePage R (eds) Computing Science and Statistics. Springer, New York, NY
21. Nagatsuka H, Balakrishnan N, Kamakura T (2013) A consistent method of estimation for the three-parameter Weibull distribution. Comp Statist Data Anal 58:210–226
22. Nagatsuka H, Balakrishnan N, Kamakura T (2014) A consistent method of estimation for the three-parameter gamma distribution. Commun Statist Theor Meth 43:3905–3926
23. Ogata Y (ed) (2015) Special topic issue: Statistical seismology research for earthquake predictability. Proc Inst Statist Math 63(1): (in Japanese)
24. Sibuya M (1960) Bivariate extreme statistics, I. Ann Inst Statist Math 11:195–210
25. Sibuya M, Takahashi R (2012) Analysis of tsunami magnitude data set. Extended abstract. Inst Statist Math Cooperative Research Report 299:50–60
26. Smith RL (1985) Maximum likelihood estimation in a class of nonregular cases. Biometrika 72:67–90
27. Wang JZ (2005) A note on estimation in the four-parameter beta distribution. Commun Statist Simul Comp 34:495–501
28. Woodroofe M (1972) Maximum likelihood estimation of a transition parameter of a truncated distribution. Ann Math Statist 43:113–122
29. Woodroofe M (1974) Maximum likelihood estimation of transition parameter of truncated distribution II. Ann Statist 2:474–488

Chapter 2
Resampling Procedures with Empirical Beta Copulas

Anna Kiriliouk, Johan Segers, and Hideatsu Tsukahara

Abstract The empirical beta copula is a simple but effective smoother of the empirical copula. Because it is a genuine copula, from which it is particularly easy to sample, it is reasonable to expect that resampling procedures based on the empirical beta copula are expedient and accurate. In this paper, after reviewing the literature on some bootstrap approximations for the empirical copula process, we first show the asymptotic equivalence of several bootstrapped processes related to the empirical and empirical beta copulas. Then we investigate the finite-sample properties of resampling schemes based on the empirical (beta) copula by the Monte Carlo simulation. More specifically, we consider interval estimation for functionals such as the rank correlation coefficients and dependence parameters of several well-known families of copulas. Here, we construct confidence intervals using several methods and compare their accuracy and efficiency. We also compute the actual size and power of symmetry tests based on several resampling schemes for the empirical and empirical beta copulas.

Keywords Bootstrap approximation · Copula · Empirical beta copula · Empirical copula · Rank correlations · Resampling · Semiparametric estimation · Test of symmetry

---

A. Kiriliouk
Faculté des sciences économiques, sociales et de gestion, Université de Namur, Rue de Bruxelles 61, B-5000 Namur, Belgium
e-mail: anna.kiriliouk@unamur.be

J. Segers
Institut de Statistique, Biostatistique et Sciences Actuarielles, Université catholique de Louvain, Voie du Roman Pays 20, B-1348 Louvain-la-Neuve, Belgium
e-mail: johan.segers@uclouvain.be

H. Tsukahara (✉)
Faculty of Economics, Seijo University, 6–1–20 Seijo, Setagaya-ku, Tokyo 157-8511, Japan
e-mail: tsukahar@seijo.ac.jp

N. Hoshino et al. (eds.), *Pioneering Works on Extreme Value Theory*, JSS Research Series in Statistics,
https://doi.org/10.1007/978-981-16-0768-4_2

2.1 Introduction

Let $\boldsymbol{X}_i = (X_{i1}, \ldots, X_{id})$, $i \in \{1, \ldots, n\}$, be independent and identically distributed (i.i.d.) random vectors, and assume that the cumulative distribution function, F, of $\boldsymbol{X}_i$ is continuous. By Sklar's theorem [20], there exists a unique copula, C, such that

$$F(\boldsymbol{x}) = C\big(F_1(x_1), \ldots, F_d(x_d)\big), \quad \boldsymbol{x} = (x_1, \ldots, x_d) \in \mathbb{R}^d,$$

where F_j is the jth marginal distribution function of F. In fact, in the continuous case, we have $C(\boldsymbol{u}) = F\big(F_1^{-}(u_1), \ldots, F_d^{-}(u_d)\big)$ for $\boldsymbol{u} = (u_1, \ldots, u_d) \in [0, 1]^d$, where $H^{-}(u) = \inf\{t \in \mathbb{R}\colon H(t) \geqslant u\}$ is the generalized inverse of a distribution function H. The empirical copula $\mathbb{C}_n$ [3] is defined by

$$\mathbb{C}_n(\boldsymbol{u}) := \mathbb{F}_n\big(\mathbb{F}_{n1}^{-}(u_1), \ldots, \mathbb{F}_{nd}^{-}(u_d)\big),$$

where, for $j \in \{1, \ldots, d\}$,

$$\mathbb{F}_n(\boldsymbol{x}) := \frac{1}{n}\sum_{i=1}^{n} \mathbb{1}\{X_{i1} \leqslant x_1, \ldots, X_{id} \leqslant x_d\}, \quad \mathbb{F}_{nj}(x_j) := \frac{1}{n}\sum_{i=1}^{n} \mathbb{1}\left\{X_{ij} \leqslant x_j\right\}.$$

For $i \in \{1, \ldots, n\}$ and $j \in \{1, \ldots, d\}$, let $R_{ij,n}$ be the rank of X_{ij} among $X_{1j}, \ldots, X_{nj}$; namely,

$$R_{ij,n} = \sum_{k=1}^{n} \mathbb{1}\{X_{kj} \leqslant X_{ij}\}. \tag{2.1}$$

A frequently used rank-based version of the empirical copula is given by

$$\tilde{\mathbb{C}}_n(\boldsymbol{u}) := \frac{1}{n}\sum_{i=1}^{n}\prod_{j=1}^{d} \mathbb{1}\left\{\frac{R_{ij,n}}{n} \leqslant u_j\right\}. \tag{2.2}$$

In the absence of ties, we have

$$\|\tilde{\mathbb{C}}_n - \mathbb{C}_n\|_\infty := \sup_{\boldsymbol{u}\in[0,1]^d} |\tilde{\mathbb{C}}_n(\boldsymbol{u}) - \mathbb{C}_n(\boldsymbol{u})| \leqslant \frac{d}{n}. \tag{2.3}$$

The functions $\mathbb{C}_n$ and $\tilde{\mathbb{C}}_n$ are both piecewise constant and cannot be genuine copulas. When the sample size is small, they suffer from the presence of ties when used in resampling.

The empirical beta copula [18], defined in Sect. 2.3, is a simple but effective way of correcting and smoothing the empirical copula. Even though its asymptotic distribution is the same as that of the usual empirical copula, its accuracy in small samples is usually better, partly because it is itself always a genuine copula. Moreover, drawing random samples from the empirical beta copula is quite straightforward.

Because of these properties, it is reasonable to expect that simple and accurate resampling schemes for the empirical copula process can be constructed based on the empirical beta copula. For tail copulas, which are limit functions describing the asymptotic behavior of a copula in the corner of the unit cube, a simulation study in [12] shows that a bootstrap based on the empirical beta copula performs significantly better than the direct multiplier bootstrap of [1]. The purpose of this study is to further investigate the finite-sample and the asymptotic behavior of this resampling method for general copulas.

The paper is structured as follows. In Sect. 2.2, we review and discuss the literature on resampling methods for the empirical copula process. The asymptotic properties of two resampling procedures based on the empirical beta copula are investigated in Sect. 2.3. In Sect. 2.4, extensive simulation studies are conducted to demonstrate the effectiveness of resampling procedures based on the empirical beta copula to construct confidence intervals for several copula functionals and to test the shape constraints on the copula. We conclude the paper with some discussion and open questions in Sect. 2.5. All proofs are relegated to the appendix.

2.2 Review on Bootstrapping Empirical Copula Processes

In this section, we give a short review on bootstrapping empirical copula processes, incorporating several newer improvements. We limit ourselves to i.i.d. sequences. Note that extensions to stationary time series are considered in [2], among others.

First, we recall a basic result on the weak convergence of the empirical copula process. Let $\ell^\infty([0,1]^d)$ be the Banach space of real-valued, bounded functions on $[0,1]^d$, equipped with the supremum norm $\|\cdot\|_\infty$. The arrow $\rightsquigarrow$ denotes weak convergence in the sense used in [23]. The following is the only condition needed for our convergence results.

Condition 2.2.1 For each $j \in \{1,\dots,d\}$, the copula C has a continuous first-order partial derivative $\dot{C}_j(\boldsymbol{u}) = \partial C(\boldsymbol{u})/\partial u_j$ on the set $\{\boldsymbol{u} \in [0,1]^d : 0 < u_j < 1\}$.

The next theorem is proved in [17]. Let $\mathbb{U}^C$ denote a C-pinned Brownian sheet, that is, a centered Gaussian process on $[0,1]^d$ with continuous trajectories and covariance function

$$\operatorname{Cov}\{\mathbb{U}^C(\boldsymbol{u}), \mathbb{U}^C(\boldsymbol{v})\} = C(\boldsymbol{u} \wedge \boldsymbol{v}) - C(\boldsymbol{u})\,C(\boldsymbol{v}), \qquad \boldsymbol{u}, \boldsymbol{v} \in [0,1]^d. \tag{2.4}$$

Theorem 2.2.2 *Suppose Condition 2.2.1 holds. Then we have*

$$\mathbb{G}_n := \sqrt{n}(\mathbb{C}_n - C) \rightsquigarrow \mathbb{G}^C, \qquad n \to \infty,$$

in $\ell^\infty([0,1]^d)$, *where*

$$\mathbb{G}^C(\boldsymbol{u}) := \mathbb{U}^C(\boldsymbol{u}) - \sum_{j=1}^{d} \dot{C}_j(\boldsymbol{u})\, \mathbb{U}^C(\mathbf{1}, u_j, \mathbf{1}),$$

with u_j appearing at the jth coordinate.

Next, we introduce notation for the convergence of conditional laws in probability given the data as defined in [14]; see also [23, Sect. 2.9]. Let

$$\mathrm{BL}_1 := \{h \colon \ell^\infty([0,1]^d) \to \mathbb{R} \mid \|h\|_\infty \leqslant 1 \text{ and } |h(x) - h(y)| \leqslant \|x - y\|_\infty \text{ for all } x, y \in \ell^\infty([0,1]^d)\}. \tag{2.5}$$

If $\hat{X}_n$ is a sequence of bootstrapped processes in $(\ell^\infty([0,1]^d), \|\cdot\|_\infty)$ with random weights W, then the notation

$$\hat{X}_n \underset{W}{\overset{\mathrm{P}}{\rightsquigarrow}} X, \qquad n \to \infty \tag{2.6}$$

means that

$$\left.\begin{aligned} &\sup_{h \in \mathrm{BL}_1} |\mathrm{E}_W[h(\hat{X}_n)] - \mathrm{E}[h(X)]| \longrightarrow 0 \quad \text{in outer probability,} \\ &\mathrm{E}_W[h(\hat{X}_n)^*] - \mathrm{E}_W[h(\hat{X}_n)_*] \overset{\mathrm{P}}{\longrightarrow} 0 \quad \text{for all } h \in \mathrm{BL}_1\,. \end{aligned}\right\} \tag{2.7}$$

Here, the notation E_W indicates conditional expectation over the weights W given the data $\boldsymbol{X}_1, \ldots, \boldsymbol{X}_n$, and $h(\hat{X}_n)^*$ and $h(\hat{X}_n)_*$ denote the minimal measurable majorant and maximal measurable minorant, respectively, with respect to the joint data $\boldsymbol{X}_1, \ldots, \boldsymbol{X}_n, W$.

In the following, the random weights W can signify different things: a multinomial random vector when drawing from the data with replacement; i.i.d. multipliers in the multiplier bootstrap; or vectors of order statistics from the uniform distribution when resampling from the empirical beta copula. In (2.6), the symbol W will then be changed accordingly.

2.2.1 Straightforward Bootstrap

Let $(W_{n1}, \ldots, W_{nn})$ be a multinomial random vector with probabilities $(1/n, \ldots, 1/n)$, independent of the sample $\boldsymbol{X}_1, \ldots, \boldsymbol{X}_n$. Set

$$\mathbb{C}_n^*(\boldsymbol{u}) = \mathbb{F}_n^*\big(\mathbb{F}_{n1}^{*-}(u_1), \ldots, \mathbb{F}_{nd}^{*-}(u_d)\big),$$

where

$$\mathbb{F}_n^*(\boldsymbol{x}) := \frac{1}{n}\sum_{i=1}^n W_{ni}\prod_{j=1}^d \mathbb{1}\{X_{ij} \leqslant x_j\},$$

$$\mathbb{F}_{nj}^*(x_j) := \frac{1}{n}\sum_{i=1}^n W_{ni}\,\mathbb{1}\{X_{ij} \leqslant x_j\}, \quad j \in \{1, \ldots, d\}.$$

We can also define the bootstrapped version of the rank-based empirical copula

$$\tilde{\mathbb{C}}_n^*(\boldsymbol{u}) = \frac{1}{n}\sum_{i=1}^n W_{ni}\prod_{j=1}^d \mathbb{1}\left\{\frac{R_{ij,n}^*}{n} \leqslant u_j\right\}, \tag{2.8}$$

where

$$R_{ij,n}^* = \sum_{k=1}^n W_{nk}\,\mathbb{1}\left\{X_{kj} \leqslant X_{ij}\right\}. \tag{2.9}$$

Since a bootstrap sample will have ties with a (large) positive probability, the bound (2.3) is no longer valid for $\mathbb{C}_n^*$ and $\tilde{\mathbb{C}}_n^*$. However, we can prove the following.

Proposition 2.2.3

$$\|\mathbb{C}_n^* - \tilde{\mathbb{C}}_n^*\|_\infty = O_p\big(n^{-1}\log n\big), \qquad n \to \infty. \tag{2.10}$$

The proof of Proposition 2.2.3 is given in the appendix. Convergence in probability of the conditional laws

$$\sqrt{n}(\mathbb{C}_n^* - \mathbb{C}_n) \underset{W}{\overset{\mathrm{P}}{\rightsquigarrow}} \mathbb{G}^C, \quad n \to \infty,$$

in the space $\ell^\infty([0,1]^d)$ is shown in [5] under the condition that all partial derivatives $\dot{C}_j$ exist and are continuous on $[0,1]^d$, and in [2] under the weaker Condition 2.2.1. From (2.3) and Proposition 2.2.3, we also have

$$\tilde{\alpha}_n := \sqrt{n}(\tilde{\mathbb{C}}_n^* - \tilde{\mathbb{C}}_n) \underset{W}{\overset{\mathrm{P}}{\rightsquigarrow}} \mathbb{G}^C, \quad n \to \infty. \tag{2.11}$$

2.2.2 *Multiplier Bootstrap with Estimated Partial Derivatives*

The multiplier bootstrap for the empirical copula, proposed by [16], has proved useful in many problems. In [1], this method is found to exhibit better finite-sample performance than other resampling methods for the empirical copula process. Here, we present a modified version proposed by [1], which we employ in the simulation studies in Sect. 2.4.

Let $\xi_1, \ldots, \xi_n$ be i.i.d. nonnegative random variables, independent of the data, with $\mathrm{E}(\xi_i) = \mu$, $\mathrm{Var}(\xi_i) = \tau^2 > 0$, and $\|\xi_i\|_{2,1} := \int_0^\infty \sqrt{\mathrm{P}(|\xi_i| > x)}\,\mathrm{d}x < \infty$. Put $\overline{\xi}_n := n^{-1}\sum_{i=1}^n \xi_i$, and set

$$\mathbb{C}_n^\circ(\boldsymbol{u}) := \frac{1}{n}\sum_{i=1}^n \frac{\xi_i}{\overline{\xi}_n}\prod_{j=1}^d \mathbb{1}\left\{X_{ij} \leqslant \mathbb{F}_{nj}^-(u_j)\right\},$$

$$\tilde{\mathbb{C}}_n^\circ(\boldsymbol{u}) := \frac{1}{n}\sum_{i=1}^n \frac{\xi_i}{\overline{\xi}_n}\prod_{j=1}^d \mathbb{1}\left\{\mathbb{F}_{nj}(X_{ij}) \leqslant u_j\right\}.$$

Define $\beta_n^\circ := \sqrt{n}(\mu/\tau)(\mathbb{C}_n^\circ - \mathbb{C}_n)$ and $\tilde{\beta}_n^\circ := \sqrt{n}(\mu/\tau)(\tilde{\mathbb{C}}_n^\circ - \tilde{\mathbb{C}}_n)$. Using Theorem 2.6 in [14] and the almost sure convergence $\|\mathbb{F}_{nj}^- - I\|_\infty \to 0$, where I is the identity function on [0, 1], we can show that

$$\beta_n^\circ \underset{\xi}{\overset{\mathrm{P}}{\rightsquigarrow}} \mathbb{U}^C \quad \text{and} \quad \tilde{\beta}_n^\circ \underset{\xi}{\overset{\mathrm{P}}{\rightsquigarrow}} \mathbb{U}^C, \qquad n \to \infty.$$

Hence, if $\hat{\dot{C}}_j(\boldsymbol{u})$ is an estimate for $\dot{C}_j(\boldsymbol{u})$, where finite differencing is applied to the empirical copula at a spacing proportional to $n^{-1/2}$, then the processes

$$\begin{cases} \alpha_n^{\mathrm{pdm}\circ}(\boldsymbol{u}) := \beta_n^\circ(\boldsymbol{u}) - \sum_{j=1}^d \hat{\dot{C}}_j(\boldsymbol{u})\,\beta_n^\circ(\mathbf{1}, u_j, \mathbf{1}) \\ \tilde{\alpha}_n^{\mathrm{pdm}\circ}(\boldsymbol{u}) := \tilde{\beta}_n^\circ(\boldsymbol{u}) - \sum_{j=1}^d \hat{\dot{C}}_j(\boldsymbol{u})\,\tilde{\beta}_n^\circ(\mathbf{1}, u_j, \mathbf{1}) \end{cases}$$

yield *conditional* approximations of $\mathbb{G}^C$, where "pdm" stands for "partial derivatives multiplier". That is, we have

$$\alpha_n^{\mathrm{pdm}\circ} \underset{\xi}{\overset{\mathrm{P}}{\rightsquigarrow}} \mathbb{G}^C \text{ and } \tilde{\alpha}_n^{\mathrm{pdm}\circ} \underset{\xi}{\overset{\mathrm{P}}{\rightsquigarrow}} \mathbb{G}^C, \qquad n \to \infty.$$

2.3 Resampling with Empirical Beta Copulas

The *empirical beta copula* [18] is defined as

$$\mathbb{C}_n^\beta(\boldsymbol{u}) = \frac{1}{n}\sum_{i=1}^n \prod_{j=1}^d F_{n,R_{ij,n}}(u_j), \qquad \boldsymbol{u} \in [0,1]^d,$$

where $R_{ij,n}$ denotes the rank, as in (2.1), and where, for $u \in [0, 1]$ and $r \in \{1, \ldots, n\}$,

$$F_{n,r}(u) = \sum_{s=r}^n \binom{n}{s} u^s (1-u)^{n-s} \tag{2.12}$$

is the cumulative distribution function of the beta distribution $\mathcal{B}(r, n+1-r)$. Note that $\mathrm{P}(U \leqslant u) = \mathrm{P}(S \geqslant r)$, for $U \sim \mathcal{B}(r, n+1-r)$ and $S \sim \mathrm{Bin}(n, u)$. In this section, we examine the asymptotic properties of two resampling procedures based on the empirical beta copula.

2.3.1 Standard Bootstrap for the Empirical Beta Copula

Let $(W_{n1}, \dots, W_{nn})$ be a multinomial random vector with success probabilities $(1/n, \dots, 1/n)$, independent of the original sample. Set

$$\mathbb{C}_n^{\beta *}(\boldsymbol{u}) = \frac{1}{n} \sum_{i=1}^{n} W_{ni} \prod_{j=1}^{d} F_{n, R_{ij,n}^*}(u_j),$$

where $R_{ij,n}^*$ are the bootstrapped ranks in (2.9). Let $S_j \sim \mathrm{Bin}(n, u_j)$, for $j = 1, \dots, d$, be d independent binomial random variables. Let E_S denote the expectation with respect to $(S_1, \dots, S_d)$, conditional on the sample and the multinomial random vector. It then follows that

$$\mathbb{C}_n^{\beta *}(\boldsymbol{u}) = \frac{1}{n} \sum_{i=1}^{n} W_{ni} \prod_{j=1}^{d} \mathrm{E}_S \left[\mathbb{1} \left\{ \frac{R_{ij,n}^*}{n} \leqslant \frac{S_j}{n} \right\} \right] = \mathrm{E}_S \left[\tilde{\mathbb{C}}_n^*(S_1/n, \dots, S_d/n) \right],$$

where $\tilde{\mathbb{C}}_n^*$ is the bootstrapped rank-based empirical copula in (2.8). Similarly, the empirical beta copula is

$$\mathbb{C}_n^{\beta}(\boldsymbol{u}) = \frac{1}{n} \sum_{i=1}^{n} \prod_{j=1}^{d} F_{n, R_{ij,n}}(u_j) = \mathrm{E}_S \left[\tilde{\mathbb{C}}_n(S_1/n, \dots, S_d/n) \right],$$

where $\tilde{\mathbb{C}}_n$ is the rank-based empirical copula in (2.2). Consider the bootstrapped processes $\tilde{\alpha}_n$ defined in (2.11) and $\alpha_n^{\beta} := \sqrt{n}(\mathbb{C}_n^{\beta *} - \mathbb{C}_n^{\beta})$. We find

$$\alpha_n^{\beta}(\boldsymbol{u}) = \mathrm{E}_S[\tilde{\alpha}_n(S_1/n, \dots, S_d/n)]. \tag{2.13}$$

Using the weak convergence of the bootstrapped process $\tilde{\alpha}_n$, we prove the following proposition. As a result, the consistency of the bootstrapped process $\tilde{\alpha}_n$ of the (rank-based) empirical copula in (2.11) entails consistency of the one for the empirical beta copula.

Proposition 2.3.1 *Under Condition 2.2.1, we have*

$$\sup_{\boldsymbol{u}\in[0,1]^d} |\alpha_n^\beta(\boldsymbol{u}) - \tilde{\alpha}_n(\boldsymbol{u})| = o_p(1), \qquad n \to \infty, \tag{2.14}$$

and thus $\alpha_n^\beta \underset{W}{\overset{P}{\rightsquigarrow}} \mathbb{G}^C$ *as* $n \to \infty$.

2.3.2 Bootstrap by Drawing Samples from the Empirical Beta Copula

The original motivation of [18] was resampling; the uniform random variables generated independently and rearranged in the order specified by the componentwise ranks of the original sample might, in some sense, be considered a bootstrap sample. Although this idea turned out to be not entirely correct, it still led to the discovery of the empirical beta copula. In the same spirit, it is natural to study the bootstrap method using samples drawn from the empirical beta copula $\mathbb{C}_n^\beta$.

It is in fact very simple to generate a random variate $\boldsymbol{V}$ from $\mathbb{C}_n^\beta$.

Algorithm 2.3.2 Given the ranks $R_{ij,n} = r_{ij},\ j = 1, \ldots, d$, of the original sample:

1. Generate I from the discrete uniform distribution on $\{1, \ldots, n\}$.
2. Independently generate $V_j^\# \sim \mathcal{B}(r_{Ij}, n + 1 - r_{Ij}),\ j \in \{1, \ldots, d\}$.
3. Set $\boldsymbol{V}^\# = (V_1^\#, \ldots, V_d^\#)$.

Repeating the above algorithm n times independently, we get a sample of n independent random vectors drawn from $\mathbb{C}_n^\beta$, conditional on the data $\boldsymbol{X}_1, \ldots, \boldsymbol{X}_n$. Let this sample be denoted by $\boldsymbol{V}_i^\# = (V_{i1}^\#, \ldots, V_{id}^\#),\ i = 1, \ldots, n$. This procedure can be viewed as a kind of *smoothed bootstrap* (see [4], [19, Sect. 3.5]) because the empirical beta copula may be thought of as a smoothed version of the empirical copula.

The joint and marginal empirical distribution functions of the bootstrap sample are

$$\mathbb{G}_n^\#(\boldsymbol{u}) = \frac{1}{n}\sum_{i=1}^n \prod_{j=1}^d \mathbb{1}\{V_{ij}^\# \leqslant u_j\} \quad \text{and} \quad \mathbb{G}_{nj}^\#(u_j) = \frac{1}{n}\sum_{i=1}^n \mathbb{1}\{V_{ij}^\# \leqslant u_j\},$$

respectively. The ranks of the bootstrap sample are given by

$$R_{ij,n}^\# = n\,\mathbb{G}_{nj}^\#(V_{ij}^\#) = \sum_{k=1}^n \mathbb{1}\{V_{kj}^\# \leqslant V_{ij}^\#\}. \tag{2.15}$$

These yield bootstrapped versions of the Deheuvels empirical copula [3], rank-based empirical copula (2.2), and empirical beta copula:

$$\mathbb{C}_n^\#(\boldsymbol{u}) := \mathbb{G}_n^\#(\mathbb{G}_{n1}^{\#-}(u_1), \ldots, \mathbb{G}_{nd}^{\#-}(u_d)), \quad \tilde{\mathbb{C}}_n^\#(\boldsymbol{u}) := \frac{1}{n}\sum_{i=1}^n \prod_{j=1}^d \mathbb{1}\{R_{ij,n}^\#/n \leqslant u_j\},$$

$$\mathbb{C}_n^{\beta\#}(\boldsymbol{u}) := \frac{1}{n}\sum_{i=1}^n \prod_{j=1}^d F_{n,R_{ij,n}^\#}(u_j),$$

respectively.

Proposition 2.3.3 *Assume Condition 2.2.1. Then, as $n \to \infty$, we have conditional weak convergence in probability, as defined in (2.6), with respect to the random vectors $\boldsymbol{V}_1^\#, \ldots, \boldsymbol{V}_n^\#$ of the bootstrapped empirical copula processes*

$$\alpha_n^\# := \sqrt{n}(\mathbb{C}_n^\# - \mathbb{C}_n), \quad \tilde{\alpha}_n^\# := \sqrt{n}(\tilde{\mathbb{C}}_n^\# - \tilde{\mathbb{C}}_n), \quad \alpha_n^{\beta\#} := \sqrt{n}(\mathbb{C}_n^{\beta\#} - \mathbb{C}_n^\beta),$$

to the limit process $\mathbb{G}^C$ defined in Theorem 2.2.2.

2.3.3 Approximating the Sampling Distributions of Rank Statistics by Resampling from the Empirical Beta Copula

Statistical inference for C often involves rank statistics. One way to justify this is to appeal to the invariance of C under coordinatewise, continuous, strictly increasing transformations. Hence, we consider a rank statistic $T(\boldsymbol{R}_1, \ldots, \boldsymbol{R}_n)$, where $\boldsymbol{R}_i := (R_{i1,n}, \ldots, R_{id,n})$ is a vector of the coordinatewise ranks of $\boldsymbol{X}_i$. Below, we suggest a way of approximating its distribution by drawing a sample from $\mathbb{C}_n^\beta$, and then computing the "bootstrap replicates". This also avoids problems with the ties encountered when drawing with replacement from the original data. The procedure is as follows.

Algorithm 2.3.4 (*Smoothed beta bootstrap*) Given $\boldsymbol{R}_1, \ldots, \boldsymbol{R}_n$:

1. Apply Algorithm 2.3.2 n times independently to obtain a bootstrap sample $\boldsymbol{V}_1^\#, \ldots, \boldsymbol{V}_n^\#$ drawn from $\mathbb{C}_n^\beta$, compute their ranks $\boldsymbol{R}_1^\#, \ldots, \boldsymbol{R}_n^\#$ as in (2.15), and put $T^\# := T(\boldsymbol{R}_1^\#, \ldots, \boldsymbol{R}_n^\#)$.
2. Repeat Step 1 a moderate-to-large number of times, B, to obtain the bootstrap replicates $T_1^\#, \ldots, T_B^\#$.
3. Use $T_1^\#, \ldots, T_B^\#$ to approximate the sampling distribution of $T(\boldsymbol{R}_1, \ldots, \boldsymbol{R}_n)$.

The validity of this procedure follows from our claim in the preceding subsection. Because the related empirical copula processes are all asymptotically equivalent, we need to examine the small-sample performance of the methods. In Sect. 2.4.2, we construct confidence intervals for several copula functionals using popular rank statistics.

2.4 Simulation Studies

We assess the performance of the bootstrap methods presented in Sects. 2.2 and 2.3 in a wide range of applications. In all of the experiments below, the number of Monte Carlo runs and the number of bootstrap replications are both set to 1000. We use Clayton, Gumbel–Hougaard, Frank, and Gauss copula families; see, for example, [15]. Most simulations are performed in R using the package copula [9]; however, the simulation described in Sect. 2.4.2 uses MATLAB.

2.4.1 *Covariance of the Limiting Process*

We compare the estimated covariances of the limiting process $\mathbb{G}^C$ based on the standard and smoothed beta bootstrap methods with those of the partial derivatives multiplier method. In [1], the latter is shown to outperform the straightforward bootstrap and the direct multiplier method. We follow the setup in [1], evaluating the covariance at four points $\{(i/3, j/3)\}$ for $i, j \in \{1, 2\}$ in the unit square. The variables $\xi_1, \ldots, \xi_n$ for the partial derivatives multiplier method are such that $\mathbb{P}[\xi_i = 0] = \mathbb{P}[\xi_i = 2] = 1/2$ for $i \in \{1, \ldots, n\}$. For the bivariate Clayton copula with parameter $\theta = 1$, Table 2.1 shows the mean squared error of the estimated covariance based on the partial derivatives multiplier method $\alpha_n^{\mathrm{pdm}\circ}$, standard beta bootstrap α_n^{β}, and smoothed beta bootstrap $\alpha_n^{\beta\#}$, for $n = 100$ and $n = 200$. The results for $\alpha_n^{\mathrm{pdm}\circ}$ are copied from Tables 3 and 4 in [1]. Both methods based on the empirical beta copula outperform the multiplier method for all points other than $(1/3, 1/3)$ and $(2/3, 2/3)$.

2.4.2 *Confidence Intervals for Rank Correlation Coefficients*

Here, we assess the performance of the straightforward bootstrap and the smoothed beta bootstrap (Sects. 2.2.1 and 2.3.3) for constructing confidence intervals for two popular rank correlation coefficients for bivariate distributions, namely, Kendall's τ and Spearman's ρ, which are known to depend only on the copula C associated with F.

Table 2.1 Mean squared error ($\times 10^4$) of the covariance estimates for the bivariate Clayton copula with $\theta = 1$

		$n = 100$				$n = 200$			
		$(\frac{1}{3}, \frac{1}{3})$	$(\frac{1}{3}, \frac{2}{3})$	$(\frac{2}{3}, \frac{1}{3})$	$(\frac{2}{3}, \frac{2}{3})$	$(\frac{1}{3}, \frac{1}{3})$	$(\frac{1}{3}, \frac{2}{3})$	$(\frac{2}{3}, \frac{1}{3})$	$(\frac{2}{3}, \frac{2}{3})$
α_n^{pdmo}	(1/3, 1/3)	0.8887	0.5210	0.5222	0.3716	0.4595	0.2673	0.2798	0.1961
	(1/3, 2/3)		1.0112	0.1799	0.2988		0.5211	0.1069	0.1577
	(2/3, 1/3)			0.9899	0.2818			0.5092	0.1681
	(2/3, 2/3)				0.6250				0.2992
α_n^{β}	(1/3, 1/3)	0.9992	0.3402	0.3473	0.1956	0.6205	0.2427	0.2383	0.1547
	(1/3, 2/3)		0.7887	0.1294	0.1889		0.4933	0.0857	0.1366
	(2/3, 1/3)			0.7644	0.1821			0.4898	0.1376
	(2/3, 2/3)				0.7108				0.4183
$\alpha_n^{\beta\#}$	(1/3, 1/3)	1.2248	0.2929	0.2924	0.1456	0.6761	0.1874	0.1888	0.1128
	(1/3, 2/3)		0.8461	0.0992	0.1691		0.4814	0.0703	0.1071
	(2/3, 1/3)			0.8856	0.1682			0.4956	0.1149
	(2/3, 2/3)				1.1209				0.5913

The population Kendall's τ is defined by

$$\tau(C) := 4 \int_0^1 \int_0^1 C(u_1, u_2)\, \mathrm{d}C(u_1, u_2) - 1.$$

In terms of

$$Q_{k,i} := \mathrm{sign}[(X_{k,1} - X_{i,1})(X_{k,2} - X_{i,2})] = \mathrm{sign}[(R_{k1,n} - R_{i1,n})(R_{k2,n} - R_{i2,n})],$$

and

$$K := \sum_{i=1}^{n-1} \sum_{k=i+1}^{n} Q_{k,i},$$

the sample Kendall's τ is given by $\hat{\tau} := 2K/[n(n-1)]$. Its asymptotic variance can be estimated by

$$\hat{\sigma}_\tau^2 := \frac{2}{n(n-1)} \left[\frac{2(n-2)}{n(n-1)^2} \sum_{i=1}^{n} (C_i - \overline{C})^2 + 1 - \hat{\tau}^2 \right],$$

where $C_i := \sum_{k=1,\, k \neq i}^{n} Q_{k,i}$, $i \in \{1, \ldots, n\}$ and $\overline{C} = n^{-1} \sum_{i=1}^{n} C_i = 2K/n$ (see [10]). Thus, an asymptotic confidence interval for τ is given by $\hat{\tau} \pm z_{\alpha/2} \hat{\sigma}_\tau$, where $z_{\alpha/2}$ is the usual standard normal tail quantile.

This interval can be compared with the confidence intervals obtained using our resampling methods. Table 2.2 shows the coverage probabilities and the average lengths of the estimated confidence intervals based on the asymptotic distribution, straightforward bootstrap, and smoothed beta bootstrap for the independence copula ($\tau = 0$) and the Clayton copula with $\theta = 2$ ($\tau = 0.5$) and $\theta = -2/3$ ($\tau = -0.5$). The nominal confidence level is 0.95. The smoothed beta bootstrap gives the most conservative coverage probabilities, but has the shortest length of the three.

The population Spearman's ρ and the sample Spearman's ρ are given by

$$\rho(C) := 12 \int_0^1 \int_0^1 \left[C(u_1, u_2) - u_1 u_2 \right] du_1 du_2,$$

$$\hat{\rho} := \frac{12}{n(n^2-1)} \sum_{i=1}^{n} \left(R_{i1,n} - \frac{n+1}{2} \right) \left(R_{i2,n} - \frac{n+1}{2} \right),$$

respectively. The limiting distribution of $\hat{\rho}$ is equal to that of $12 \iint \mathbb{G}^C(u_1, u_2) \mathrm{d}u_1 \mathrm{d}u_2$; thus, in principle, it is possible to construct confidence intervals based on the asymptotics. However, unlike the case of $\hat{\tau}$, this procedure is cumbersome and involves partial derivatives of C, which must be estimated. Therefore, we omit it from our study. We continue to set the nominal confidence level to 0.95 in the experiment. Table 2.3 shows that the coverage probabilities for the smoothed beta bootstrap are more conservative than those for the straightforward bootstrap; however, the average

Table 2.2 Coverage probabilities and average lengths of the confidence intervals for Kendall's τ for the Clayton copula family, computed using the normal approximation, straightforward bootstrap, and smoothed beta bootstrap

		$\tau = 0$				$\tau = 0.5$				$\tau = -0.5$			
	n	40	60	80	100	40	60	80	100	40	60	80	100
Coverage probability	asymp	0.952	0.930	0.941	0.959	0.946	0.931	0.937	0.943	0.933	0.941	0.939	0.926
	boot	0.957	0.937	0.942	0.963	0.949	0.940	0.949	0.949	0.951	0.947	0.938	0.935
	beta	0.964	0.949	0.949	0.966	0.952	0.947	0.954	0.955	0.963	0.935	0.948	0.939
Average length	asymp	0.449	0.355	0.304	0.271	0.364	0.287	0.245	0.217	0.378	0.302	0.257	0.227
	boot	0.450	0.357	0.306	0.272	0.366	0.288	0.246	0.218	0.380	0.304	0.258	0.228
	beta	0.433	0.347	0.299	0.268	0.350	0.279	0.240	0.213	0.365	0.294	0.253	0.224

Table 2.3 Coverage probabilities and average lengths of the confidence intervals for Spearman's ρ for the Clayton copula family, based on the straightforward bootstrap and smoothed beta bootstrap

		$\rho = 0$				$\rho = 0.5$				$\rho = -0.5$			
	n	40	60	80	100	40	60	80	100	40	60	80	100
Coverage probability	boot	0.956	0.943	0.953	0.951	0.959	0.953	0.949	0.952	0.952	0.954	0.960	0.956
	beta	0.965	0.946	0.957	0.956	0.961	0.958	0.960	0.952	0.969	0.957	0.964	0.958
Average length	boot	0.634	0.514	0.444	0.397	0.524	0.424	0.367	0.326	0.519	0.418	0.366	0.324
	beta	0.625	0.510	0.442	0.395	0.522	0.424	0.368	0.325	0.519	0.418	0.367	0.324

lengths of the estimated confidence intervals are very similar in the two methods. This could be due to the fact that $\rho(\mathbb{C}_n^\beta) = [(n-1)/(n+1)]\hat{\rho}$, as can be computed directly.

2.4.3 *Confidence Intervals for a Copula Parameter*

Suppose that the copula of F is parametrized by $\theta \in \Theta \subset \mathbb{R}$, such that $F(x_1, x_2) = C_\theta(F_1(x_1), F_2(x_2))$. When the F_j's are unknown, the resulting problem of estimating θ is semiparametric; see [6, 21]. Assume that C_θ is absolutely continuous with density c_θ, which is differentiable with respect to θ. Replacing the unknown F_j's in the score equation with their (rescaled) empirical counterparts, one gets the estimating equation

$$\sum_{k=1}^{n} \frac{\dot{c}_\theta[\mathbb{F}_{n1}(X_{k,1}), \mathbb{F}_{n2}(X_{k,2})]}{c_\theta[\mathbb{F}_{n1}(X_{k,1}), \mathbb{F}_{n2}(X_{k,2})]} = 0, \tag{2.16}$$

where $\dot{c}_\theta = \partial c_\theta / \partial \theta$. The solution $\widehat{\theta}$ to (2.16) is called the *pseudo-likelihood estimator*.

We compare the confidence intervals for θ estimated using the pseudo-likelihood estimator $\widehat{\theta}$ based on the asymptotic variance given in [6], straightforward bootstrap, smoothed beta bootstrap, and classic parametric bootstrap. We set the nominal confidence level equal to 0.95. Tables 2.4 and 2.5 show the estimated coverage probabilities and average interval lengths of the confidence intervals for the Clayton, Gauss, Frank, and Gumbel–Hougaard copula families, respectively. For the Clayton copula, the smoothed beta bootstrap gives the shortest intervals, both for $\theta = 1$ and $\theta = 2$, but for $\theta = 2$, the coverage probabilities are too liberal, which is somewhat puzzling. For the Frank and Gumbel–Hougaard copulas, the smoothed beta bootstrap gives the most conservative coverage probabilities, but has the shortest length of the four. For the Gauss copula, the asymptotic approximation gives significantly smaller coverage probabilities than the nominal value of 0.95.

2.4.4 *Testing the Symmetry of a Copula*

For a bivariate copula C, consider the problem of testing the symmetry hypothesis $H_0 : C(u_1, u_2) = C(u_2, u_1)$ for all $(u_1, u_2) \in [0, 1]^2$. We focus on the following two test statistics proposed in [7]:

Table 2.4 Coverage probabilities and average lengths of the confidence intervals for the parameter of the Clayton copula, with $\theta = 1$ ($\tau = 1/3$) and $\theta = 2$ ($\tau = 1/2$). Intervals are computed using the asymptotic normal approximation, straightforward bootstrap, smoothed beta bootstrap, and parametric bootstrap

		$\theta = 1$				$\theta = 2$			
	n	40	60	80	100	40	60	80	100
Coverage probability	asymp	0.954	0.969	0.960	0.965	0.951	0.940	0.940	0.946
	boot	0.953	0.943	0.944	0.943	0.968	0.952	0.953	0.951
	beta	0.953	0.964	0.957	0.952	0.933	0.904	0.908	0.906
	param	0.924	0.923	0.933	0.948	0.957	0.951	0.955	0.953
Average length	asymp	2.011	1.632	1.354	1.237	2.764	2.142	1.821	1.615
	boot	1.894	1.449	1.198	1.046	2.991	2.205	1.841	1.626
	beta	1.517	1.225	1.050	0.935	1.957	1.612	1.420	1.296
	param	1.914	1.448	1.222	1.070	2.821	2.150	1.829	1.617

Table 2.5 Coverage probabilities and average lengths of confidence intervals for the parameter of the Gaussian copula with $\theta = 1/\sqrt{2}$, the Frank copula with $\theta = 5.75$, and the Gumbel–Hougaard copula with $\theta = 2$. All copulas have $\tau \approx 1/2$. Intervals computed via the asymptotic normal approximation, the straightforward bootstrap, the smoothed beta bootstrap, and the parametric bootstrap

		Gauss				Frank				Gumbel–Hougaard			
	n	40	60	80	100	40	60	80	100	40	60	80	100
Coverage probability	asymp	0.881	0.895	0.910	0.928	0.941	0.950	0.948	0.965	0.954	0.940	0.940	0.955
	boot	0.942	0.944	0.947	0.950	0.957	0.956	0.946	0.963	0.965	0.951	0.953	0.965
	beta	0.968	0.962	0.970	0.953	0.965	0.961	0.952	0.965	0.970	0.951	0.952	0.954
	param	0.903	0.921	0.923	0.930	0.938	0.956	0.941	0.962	0.924	0.926	0.932	0.945
Average length	asymp	0.303	0.274	0.213	0.193	5.699	4.487	3.821	3.391	1.425	1.082	0.929	0.816
	boot	0.319	0.257	0.219	0.197	6.139	4.677	3.949	3.464	1.572	1.162	0.968	0.855
	beta	0.341	0.269	0.228	0.203	5.367	4.335	3.735	3.329	1.170	0.947	0.826	0.747
	param	0.292	0.242	0.210	0.191	5.729	4.494	3.848	3.389	1.546	1.170	0.983	0.869

$$S_n = \int_{[0,1]^2} [\mathbb{C}_n(u_1, u_2) - \mathbb{C}_n(u_2, u_1)]^2 \, \mathrm{d}\mathbb{C}_n(u_1, u_2),$$
$$R_n = \int_{[0,1]^2} [\mathbb{C}_n(u_1, u_2) - \mathbb{C}_n(u_2, u_1)]^2 \, \mathrm{d}u_1 \, \mathrm{d}u_2,$$

and include versions based on the empirical beta copula; that is,

$$S_n^\beta = \int_{[0,1]^2} \left[\mathbb{C}_n^\beta(u_1, u_2) - \mathbb{C}_n^\beta(u_2, u_1)\right]^2 \mathrm{d}\mathbb{C}_n^\beta(u_1, u_2),$$
$$R_n^\beta = \int_{[0,1]^2} [\mathbb{C}_n^\beta(u_1, u_2) - \mathbb{C}_n^\beta(u_2, u_1)]^2 \, \mathrm{d}u_1 \, \mathrm{d}u_2.$$

Similarly, as in Proposition 1 in [7], the statistic R_n^β can be computed as

$$R_n^\beta = \frac{2}{n^2} \sum_{i=1}^n \sum_{j=1}^n \{B_n(R_{i1,n}, R_{j1,n}) B_n(R_{i2,n}, R_{j2,n}) - B_n(R_{i1,n}, R_{j2,n}) B_n(R_{i2,n}, R_{j1,n})\},$$

with $B_n(r, s) = \int_0^1 F_{n,r}(u) F_{n,s}(u) \, \mathrm{d}u$ for $r, s \in \{1, \ldots, n\}$ and $F_{n,r}(u)$ as in (2.12). For fixed n, the matrix B_n can be precomputed and stored, which reduces the computation time for the resampling methods. Similarly, S_n^β can be written as

$$\begin{aligned} S_n^\beta = n^{-3} \sum_{i=1}^n \sum_{j=1}^n \sum_{k=1}^n \{ & C_n(R_{i1,n}, R_{j1,n} R_{k1,n}) C_n(R_{i2,n}, R_{j2,n} R_{k2,n}) \\ & - C_n(R_{i1,n}, R_{j2,n} R_{k1,n}) C_n(R_{i2,n}, R_{j1,n} R_{k2,n}) \\ & - C_n(R_{i2,n}, R_{j1,n} R_{k1,n}) C_n(R_{i1,n}, R_{j2,n} R_{k2,n}) \\ & + C_n(R_{i2,n}, R_{j2,n} R_{k1,n}) C_n(R_{i1,n}, R_{j1,n} R_{k2,n}) \}, \end{aligned}$$

with $C_n(r, s, t) = \int_0^1 F_{n,r}(u) F_{n,s}(u) \, \mathrm{d}F_{n,t}(u)$ for $r, s, t \in \{1, \ldots, n\}$.

In order to compute the p-values, we need to generate bootstrap samples from a distribution that fulfills the restriction specified by H_0. A natural candidate is a "symmetrized" version of the empirical beta copula

$$\mathbb{C}_n^{\beta,\mathrm{sym}}(u_1, u_2) := \frac{1}{2} \mathbb{C}_n^\beta(u_1, u_2) + \frac{1}{2} \mathbb{C}_n^\beta(u_2, u_1).$$

When resampling, this simply amounts to interchanging the two coordinates at random in step 3 of Algorithm 2.3.2. We employ the following three resampling schemes to compare the actual sizes of the tests.

- The symmetrized smoothed beta bootstrap: we resample from $\mathbb{C}_n^{\beta,\mathrm{sym}}$ to obtain bootstrap replicates of R_n, R_n^β, S_n, and S_n^β;

- The symmetrized version of the straightforward bootstrap for R_n and S_n;
- *exchTest* in the R package copula [9], which implements the multiplier bootstrap for R_n and S_n, as described in [7] and in Sect. 5 of [13]. For R_n, the grid length in *exchTest* is set to $m = 50$.

We use the nominal size $\alpha = 0.05$ throughout the experiment. Tables 2.6 and 2.7 show the actual sizes of the symmetry tests for the Clayton and Gauss copulas. On the whole, the smoothed beta bootstrap works better than *exchTest*, and works equally well as R_n and S_n, except when the dependence is strong ($\tau = 0.75$) and the sample size is small ($n = 50$). However, no method produces a satisfying result in the latter case. The smoothed beta bootstraps with R_n^β and S_n^β produce actual sizes similar to those based on R_n and S_n. The statistic S_n performs slightly better than R_n on average, especially for strong positive dependence. The straightforward bootstrap performs poorly in all cases, as expected [16].

To compare the power of the tests, the Clayton and Gauss copulas are made asymmetric using Khoudraji's device [11]; that is, the asymmetric version of a copula C is defined as

$$K_\delta(u_1, u_2) = u_1^\delta C(u_1^{1-\delta}, u_2), \qquad (u_1, u_2) \in [0, 1]^2.$$

Table 2.8 shows the empirical power of R_n and R_n^β for $\delta \in \{0.25, 0.5, 0.75\}$ for the three resampling methods. As shown, the smoothed beta bootstraps with R_n and R_n^β have higher power than *exchTest* for almost all sample sizes and parameter values considered; furthermore, the smoothed beta bootstrap with R_n^β has slightly higher power in almost all cases.

2.5 Concluding Remarks

We have studied the performance of resampling procedures based on the empirical beta copula and proved that all related empirical copula processes exhibit asymptotically equivalent behavior. A comparative analysis based on the Monte Carlo experiments shows that, on the whole, the smoothed beta bootstrap works fairly well, providing a useful alternative to existing resampling schemes. However, we also find that its effectiveness varies somewhat between copulas.

Higher-order asymptotics for the various nonparametric copula estimators might improve our understanding of the various resampling procedures [8, 19], although calculating such expansions seems a formidable task.

Acknowledgements The research of H. Tsukahara was supported by JSPS KAKENHI Grant Number 18H00836. The authors wish to thank an anonymous reviewer for his/her careful reading of the manuscript and helpful comments.

Table 2.6 Actual sizes of the symmetry tests based on R_n and S_n for the Clayton copula ($\theta \in \{-1/3, 2/3, 2, 6\}$), with p-values computed by the multiplier bootstrap (exchTest), straightforward bootstrap (boot), smoothed beta bootstrap (beta), and of the test based on R_n^β and S_n^β, with p-values computed by the smoothed beta bootstrap (beta2). The nominal size is $\alpha = 0.05$

	S_n						R_n				
	n	50	100	200	400		n	50	100	200	400
$\tau = -1/5$	exchTest	0.055	0.033	0.039	0.040	$\tau = -1/5$	exchTest	0.044	0.035	0.039	0.051
	boot	0.021	0.024	0.031	0.034		boot	0.009	0.019	0.027	0.044
	beta	0.057	0.038	0.039	0.041		beta	0.046	0.035	0.042	0.059
	beta2	0.042	0.036	0.043	0.057		beta2	0.050	0.037	0.041	0.059
$\tau = 0.25$	exchTest	0.039	0.029	0.036	0.036	$\tau = 0.25$	exchTest	0.030	0.022	0.040	0.031
	boot	0.009	0.015	0.026	0.034		boot	0.001	0.011	0.020	0.030
	beta	0.039	0.039	0.044	0.046		beta	0.042	0.032	0.041	0.043
	beta2	0.043	0.034	0.041	0.044		beta2	0.033	0.033	0.044	0.045
$\tau = 0.5$	exchTest	0.033	0.020	0.026	0.019	$\tau = 0.5$	exchTest	0.015	0.014	0.030	0.031
	boot	0.001	0.008	0.015	0.017		boot	0.001	0.005	0.017	0.025
	beta	0.030	0.029	0.039	0.028		beta	0.020	0.022	0.040	0.047
	beta2	0.029	0.024	0.046	0.041		beta2	0.019	0.023	0.045	0.046
$\tau = 0.75$	exchTest	0.025	0.026	0.018	0.014	$\tau = 0.75$	exchTest	0.000	0.007	0.007	0.012
	boot	0.000	0.002	0.001	0.008		boot	0.000	0.000	0.001	0.004
	beta	0.006	0.017	0.026	0.029		beta	0.000	0.006	0.017	0.029
	beta2	0.001	0.014	0.029	0.034		beta2	0.000	0.007	0.021	0.036

Table 2.7 Actual sizes of the symmetry tests based on R_n and S_n for the Gauss copula, with p-values computed by the multiplier bootstrap (exchTest), straightforward bootstrap (boot), smoothed beta bootstrap (beta), and of the test based on R_n^{β} and S_n^{β}, with p-values computed by the smoothed beta bootstrap (beta2). The nominal size is $\alpha = 0.05$

	S_n						R_n				
	n	50	100	200	400		n	50	100	200	400
$\tau = -0.5$	exchTest	0.047	0.032	0.038	0.039	$\tau = -0.5$	exchTest	0.026	0.037	0.037	0.041
	boot	0.022	0.023	0.032	0.040		boot	0.007	0.014	0.027	0.033
	beta	0.044	0.030	0.035	0.043		beta	0.020	0.030	0.036	0.040
	beta2	0.028	0.035	0.045	0.043		beta2	0.022	0.034	0.041	0.042
$\tau = 0.25$	exchTest	0.028	0.035	0.031	0.040	$\tau = 0.25$	exchTest	0.029	0.025	0.030	0.041
	boot	0.007	0.015	0.023	0.038		boot	0.008	0.015	0.025	0.037
	beta	0.033	0.038	0.039	0.045		beta	0.040	0.030	0.033	0.048
	beta2	0.048	0.033	0.037	0.047		beta2	0.037	0.031	0.037	0.048
$\tau = 0.5$	exchTest	0.034	0.018	0.025	0.029	$\tau = 0.5$	exchTest	0.011	0.014	0.019	0.034
	boot	0.003	0.005	0.015	0.026		boot	0.001	0.005	0.006	0.028
	beta	0.032	0.033	0.041	0.044		beta	0.016	0.024	0.031	0.042
	beta2	0.029	0.026	0.034	0.048		beta2	0.018	0.023	0.033	0.047
$\tau = 0.75$	exchTest	0.018	0.017	0.011	0.008	$\tau = 0.75$	exchTest	0.001	0.001	0.005	0.009
	boot	0.000	0.000	0.002	0.006		boot	0.000	0.000	0.000	0.003
	beta	0.006	0.015	0.021	0.029		beta	0.001	0.001	0.010	0.028
	beta2	0.002	0.005	0.016	0.035		beta2	0.001	0.001	0.011	0.027

Table 2.8 Actual power of the symmetry tests based on R_n, with p-values computed by the multiplier bootstrap (exchTest), straightforward bootstrap (boot), smoothed beta bootstrap (beta), and of the test based on R_n^β, with p-values computed by the smoothed beta bootstrap (beta2), for the Clayton and Gauss copulas, made asymmetric by Khoudraji's device. The nominal size is $\alpha = 0.05$

	Clayton						Gauss				
$\delta = 0.25$	n	50	100	200	400		n	50	100	200	400
$\tau = 0.25$	exchTest	0.025	0.031	0.042	0.049	$\tau = 0.25$	exchTest	0.033	0.028	0.034	0.050
	boot	0.006	0.017	0.029	0.042		boot	0.006	0.010	0.032	0.047
	beta	0.034	0.039	0.056	0.050		beta	0.044	0.042	0.048	0.057
	beta2	0.034	0.041	0.055	0.054		beta2	0.046	0.033	0.048	0.059
$\tau = 0.5$	exchTest	0.047	0.090	0.197	0.449	$\tau = 0.5$	exchTest	0.052	0.078	0.188	0.401
	boot	0.004	0.041	0.145	0.407		boot	0.007	0.028	0.136	0.366
	beta	0.060	0.100	0.216	0.469		beta	0.061	0.088	0.198	0.433
	beta2	0.062	0.103	0.227	0.486		beta2	0.065	0.098	0.212	0.441
$\tau = 0.75$	exchTest	0.199	0.630	0.985	1.000	$\tau = 0.75$	exchTest	0.205	0.637	0.973	1.000
	boot	0.038	0.380	0.949	1.000		boot	0.051	0.338	0.921	1.000
	beta	0.227	0.637	0.981	1.000		beta	0.208	0.614	0.974	1.000
	beta2	0.242	0.667	0.986	1.000		beta2	0.225	0.639	0.986	1.000
$\delta = 0.5$	n	50	100	200	400		n	50	100	200	400
$\tau = 0.25$	exchTest	0.028	0.029	0.050	0.055	$\tau = 0.25$	exchTest	0.044	0.053	0.053	0.083
	boot	0.008	0.011	0.031	0.054		boot	0.009	0.019	0.034	0.068
	beta	0.035	0.033	0.056	0.064		beta	0.051	0.059	0.055	0.090
	beta2	0.038	0.037	0.059	0.064		beta2	0.052	0.062	0.056	0.091

(continued)

Table 2.8 (continued)

	Clayton						Gauss				
$\delta = 0.25$	n	50	100	200	400		n	50	100	200	400
$\tau = 0.5$	exchTest	0.069	0.127	0.269	0.576	$\tau = 0.5$	exchTest	0.100	0.203	0.388	0.730
	boot	0.015	0.068	0.219	0.539		boot	0.027	0.119	0.326	0.695
	beta	0.077	0.140	0.299	0.593		beta	0.105	0.213	0.410	0.741
	beta2	0.075	0.144	0.306	0.602		beta2	0.116	0.225	0.416	0.750
$\tau = 0.75$	exchTest	0.385	0.814	0.997	1.000	$\tau = 0.75$	exchTest	0.478	0.914	1.000	1.000
	boot	0.125	0.644	0.993	1.000		boot	0.198	0.792	1.000	1.000
	beta	0.393	0.824	0.997	1.000		beta	0.475	0.916	1.000	1.000
	beta2	0.425	0.833	0.998	1.000		beta2	0.507	0.923	1.000	1.000
$\delta = 0.75$	n	50	100	200	400		n	50	100	200	400
$\tau = 0.25$	exchTest	0.032	0.039	0.046	0.054	$\tau = 0.25$	exchTest	0.043	0.046	0.056	0.076
	boot	0.008	0.020	0.030	0.047		boot	0.016	0.027	0.036	0.060
	beta	0.036	0.043	0.051	0.060		beta	0.047	0.053	0.061	0.081
	beta2	0.039	0.045	0.054	0.058		beta2	0.053	0.053	0.060	0.081
$\tau = 0.5$	exchTest	0.042	0.089	0.129	0.266	$\tau = 0.5$	exchTest	0.088	0.169	0.317	0.655
	boot	0.012	0.045	0.100	0.251		boot	0.030	0.113	0.271	0.604
	beta	0.053	0.094	0.132	0.279		beta	0.099	0.184	0.342	0.660
	beta2	0.053	0.102	0.140	0.287		beta2	0.102	0.194	0.354	0.670
$\tau = 0.75$	exchTest	0.144	0.372	0.693	0.962	$\tau = 0.75$	exchTest	0.369	0.636	0.947	1.000
	boot	0.051	0.268	0.645	0.958		boot	0.133	0.510	0.918	1.000
	beta	0.156	0.382	0.720	0.965		beta	0.303	0.637	0.950	1.000
	beta2	0.169	0.402	0.727	0.966		beta2	0.334	0.664	0.954	1.000

Appendix: Mathematical Proofs

Proof of Proposition 2.2.3 Let $\mathcal{N}_n = \{i = 1, \ldots, n : W_{ni} \geqslant 1\}$ be the set of indices that are sampled at least once. Then, $\mathbb{F}^*_{nj}$ is a discrete distribution function with atoms $\{X_{ij} : i \in \mathcal{N}_n\}$ and probabilities $n^{-1} W_{ni}$.

Since $n^{-1} R^*_{ij,n} = \mathbb{F}^*_{nj}(X_{ij})$, we have

$$\begin{aligned} \left|\mathbb{C}^*_n(\boldsymbol{u}) - \tilde{\mathbb{C}}^*_n(\boldsymbol{u})\right| &\leqslant \frac{1}{n} \sum_{i \in \mathcal{N}_n} W_{ni} \left| \prod_{j=1}^d \mathbb{1}\{X_{ij} \leqslant \mathbb{F}^{*-}_{nj}(u_j)\} - \prod_{j=1}^d \mathbb{1}\{\mathbb{F}^*_{nj}(X_{ij}) \leqslant u_j\} \right| \\ &\leqslant \frac{1}{n} \sum_{i \in \mathcal{N}_n} W_{ni} \sum_{j=1}^d \left| \mathbb{1}\{X_{ij} \leqslant \mathbb{F}^{*-}_{nj}(u_j)\} - \mathbb{1}\{\mathbb{F}^*_{nj}(X_{ij}) \leqslant u_j\} \right| \\ &= \frac{1}{n} \sum_{i \in \mathcal{N}_n} W_{ni} \sum_{j=1}^d \left| \mathbb{1}\{X_{ij} = \mathbb{F}^{*-}_{nj}(u_j)\} - \mathbb{1}\{\mathbb{F}^*_{nj}(X_{ij}) = u_j\} \right|. \end{aligned}$$

In the last equality, we use the fact that $x < G^-(u)$ if and only if $G(x) < u$ for any (right-continuous) distribution function G, any real x, and any $u \in [0, 1]$. For each $j \in \{1, \ldots, d\}$, $\mathbb{F}^*_{nj}(X_{ij}) = u_j$ implies $X_{ij} = \mathbb{F}^{*-}_{nj}(u_j)$ since $\mathbb{F}^*_{nj}$ jumps at X_{ij}, $i \in \mathcal{N}_n$, and there is at most a single $i \in \mathcal{N}_n$ such that $X_{ij} = \mathbb{F}^{*-}_{nj}(u_j)$. Thus, we have

$$\left|\mathbb{C}^*_n(\boldsymbol{u}) - \tilde{\mathbb{C}}^*_n(\boldsymbol{u})\right| \leqslant \frac{d}{n} \max_{i=1,\ldots,n} W_{ni}.$$

By coupling the multinomial random vector $(W_{n1}, \ldots, W_{nn})$ to a vector of independent Poisson(1) random variables $(W'_{n1}, \ldots, W'_{nn})$, as in [23, pp. 346–348], it can be shown that $\max_{i=1,\ldots,n} W_{ni} = O_p(\log n)$ as $n \to \infty$. Equation (2.10) follows. □

Proof of Proposition 2.3.1 Fix $\varepsilon > 0$. We know from (2.11) that $\tilde{\alpha}_n$ converges weakly in $\ell^\infty([0, 1]^d)$ to a Gaussian process with continuous trajectories. Write $\boldsymbol{S} = (S_1, \ldots, S_d)$ and for a point $\boldsymbol{x} \in \mathbb{R}^d$, put $|\boldsymbol{x}|_\infty = \max(|x_1|, \ldots, |x_d|)$. Furthermore, put $\|f\|_\infty = \sup\{|f(\boldsymbol{u})| : \boldsymbol{u} \in [0, 1]^d\}$ for $f : [0, 1]^d \to \mathbb{R}$. Then,

$$\begin{aligned} |\alpha^\beta_n(\boldsymbol{u}) - \tilde{\alpha}_n(\boldsymbol{u})| &\leqslant \mathrm{E}_S\left[|\tilde{\alpha}_n(S_1/n, \ldots, S_d/n) - \tilde{\alpha}_n(\boldsymbol{u})|\right] \\ &\leqslant 2\|\tilde{\alpha}_n\|_\infty \, \mathrm{P}_S[|\boldsymbol{S}/n - \boldsymbol{u}|_\infty > \varepsilon] + \sup_{\substack{\boldsymbol{v},\boldsymbol{w} \in [0,1]^d \\ |\boldsymbol{v}-\boldsymbol{w}|_\infty \leqslant \varepsilon}} |\tilde{\alpha}_n(\boldsymbol{v}) - \tilde{\alpha}_n(\boldsymbol{w})|, \end{aligned}$$

where E_S and P_S denote the expectation and probability, respectively, with respect to $\boldsymbol{S}$, conditional on the sample and the multinomial random vector. Let $Y_n(\varepsilon)$ denote the supremum on the right-hand side. By Tchebysheff's inequality, the probability in the first term on the right-hand side is bounded by a constant multiple of $n^{-1/2}\varepsilon^{-1}$ and thus tends to zero uniformly in $\boldsymbol{u} \in [0, 1]^d$. Since $\|\tilde{\alpha}_n\|_\infty = O_p(1)$, we get $\|\alpha^\beta_n - \tilde{\alpha}_n\|_\infty = o_p(1) + Y_n(\varepsilon)$ as $n \to \infty$. By the weak convergence of $\tilde{\alpha}_n$ in $\ell^\infty([0, 1])$ to a process with continuous trajectories, we can find, for any $\eta > 0$, a sufficiently small $\varepsilon > 0$ such that $\limsup_{n\to\infty} \mathrm{P}[Y_n(\varepsilon) > \eta] \leqslant \eta$. Equation (2.14) follows. □

Proof of Proposition 2.3.3 *Step 1.* Recall the C-pinned Brownian sheet $\mathbb{U}^C$ defined prior to Theorem 2.2.2. We show that

$$\gamma_n^{\#} := \sqrt{n}(\mathbb{G}_n^{\#} - \mathbb{C}_n^{\beta}) \underset{V}{\overset{\mathrm{P}}{\rightsquigarrow}} \mathbb{U}^C, \qquad n \to \infty. \tag{2.17}$$

From (2.7), two claims need to be shown: convergence in the bounded Lipschitz metric (Step 1.1), and asymptotic measurability (Step 1.2).

Step 1.1. Let $\mathcal{P}$ denote the set of all Borel probability measures on $[0, 1]^d$. For $P \in \mathcal{P}$, let $\mathbb{U}^P$ denote a tight, P-Brownian bridge on $[0, 1]^d$. Specifically, $\mathbb{U}^P$ is a centered Gaussian process with covariance function $\mathrm{E}[\mathbb{U}^P(\boldsymbol{u})\mathbb{U}^P(\boldsymbol{v})] = F(\boldsymbol{u} \wedge \boldsymbol{v}) - F(\boldsymbol{u})F(\boldsymbol{v})$, where F is the cumulative distribution function associated with P, and whose trajectories are uniformly continuous, almost surely with respect to the standard deviation semimetric [23, Example 1.5.10]:

$$d_P(\boldsymbol{u}, \boldsymbol{v}) = \big(\mathrm{E}[\{\mathbb{U}^P(\boldsymbol{u}) - \mathbb{U}^P(\boldsymbol{v})\}^2]\big)^{1/2} = \big(F(\boldsymbol{u}) - 2F(\boldsymbol{u} \wedge \boldsymbol{v}) + F(\boldsymbol{v})\big)^{1/2}. \tag{2.18}$$

Furthermore, for $P \in \mathcal{P}$, let $\mathbb{U}_{n,P}$ denote the empirical process based on independent random sampling from P. We view $\mathbb{U}_{n,P}$ as a random element of $\ell^\infty([0, 1]^d)$ from the empirical and true cumulative distribution functions. Let $P_n^\beta \in \mathcal{P}$ be the (random) probability measure associated with the empirical beta copula $\mathbb{C}_n^\beta$. Recall BL_1 in (2.5).

We need to show that, as $n \to \infty$,

$$\sup_{h\in \mathrm{BL}_1} \left| \mathrm{E}^*_{P_n^\beta}[h(\mathbb{U}_{n,P_n^\beta})] - \mathrm{E}[h(\mathbb{U}^C)] \right| \longrightarrow 0 \qquad \text{in outer probability.}$$

By the triangle inequality, it is sufficient to show the pair of convergences

$$\sup_{h\in \mathrm{BL}_1} \left| \mathrm{E}^*_{P_n^\beta}[h(\mathbb{U}_{n,P_n^\beta})] - \mathrm{E}[h(\mathbb{U}^{P_n^\beta})] \right| \longrightarrow 0, \tag{2.19}$$

$$\sup_{h\in \mathrm{BL}_1} \left| \mathrm{E}[h(\mathbb{U}^{P_n^\beta})] - \mathrm{E}[h(\mathbb{U}^C)] \right| \longrightarrow 0, \tag{2.20}$$

as $n \to \infty$, in outer probability. We do so in Steps 1.1.1 and 1.1.2, respectively.

Step 1.1.1. Identify a point $\boldsymbol{u} \in [0, 1]^d$ with the indicator function $\mathbb{1}_{(-\infty,\boldsymbol{u}]}$ on $\mathbb{R}^d$. The resulting class $\mathcal{F} = \{\mathbb{1}_{(-\infty,\boldsymbol{u}]} : \boldsymbol{u} \in [0, 1]^d\}$, being bounded (by 1) and VC [23, Example 2.6.1], it satisfies the uniform entropy condition (2.5.1) in [23]; see Theorem 2.6.7 in the same book. From their Theorem 2.8.3, we obtain the uniform Donsker property

$$\sup_{P\in\mathcal{P}} \sup_{h\in \mathrm{BL}_1} \left| \mathrm{E}^*_P[h(\mathbb{U}_{n,P})] - \mathrm{E}[h(\mathbb{U}^P)] \right| \to 0, \qquad n \to \infty. \tag{2.21}$$

The supremum over h in (2.19) is bounded by the double supremum over P and h in (2.21). The convergence in (2.19) is thus proved.

Step 1.1.2. We need to show that, almost surely, $\mathbb{U}^{P_n^\beta} \rightsquigarrow \mathbb{U}^C$ as $n \to \infty$. All processes involved are tight, centered Gaussian processes, with covariance functions determined in (2.4) using $\mathbb{C}_n^P$ or C. The strong consistency of the empirical copula, together with [18, Proposition 2.8], yields

$$\|\mathbb{C}_n^\beta - C\|_\infty \to 0, \qquad n \to \infty, \text{ a.s.} \tag{2.22}$$

This property implies (2.20). First, (2.22) implies the almost sure convergence of the covariance function of $\mathbb{U}^{P_n^\beta}$ to that of $\mathbb{U}^C$, and thus the almost sure convergence of the finite-dimensional distributions. Second, the asymptotic tightness a.s. follows from the uniform continuity of the trajectories with respect to their respective intrinsic standard deviation semimetrics (2.18) and the uniform convergence a.s. of these standard deviation semimetrics, again by (2.22).

Step 1.2. The asymptotic measurability of $\gamma_n^\#$ follows from the *unconditional* (i.e., jointly in $\boldsymbol{X}_1, \ldots, \boldsymbol{X}_n, \boldsymbol{V}_1^\#, \ldots, \boldsymbol{V}_n^\#$) weak convergence $\gamma_n^\# \rightsquigarrow \mathbb{U}^C$ as $n \to \infty$. These claims can be divided into the convergence of the finite-dimensional distributions and asymptotic tightness. The former can be shown using the Lindeberg central limit theorem for triangular arrays conditional on $\boldsymbol{X}_1, \ldots, \boldsymbol{X}_n$, using a similar method to that in the proof of Theorem 23.4 in [22]. The latter follows as in Theorems 2.5.2 and 2.8.3 in [23, p. 128 and 171], conditional on $\boldsymbol{X}_1, \ldots, \boldsymbol{X}_n$ using the fact that the class of indicator functions of cells in $\mathbb{R}^d$ is a VC-class [23, Example 2.5.4].

Step 2. Consider a map Φ that sends a cumulative distribution function H on $[0, 1]^d$ whose marginals do not assign mass at zero to the function $\boldsymbol{u} \mapsto H(H_1^-(u_1), \ldots, H_d^-(u_d))$. We have $\mathbb{C}_n^\# = \Phi(\mathbb{G}_n^\#)$ and $\mathbb{C}_n^\beta = \Phi(\mathbb{C}_n^\beta)$. By [2, Theorem 2.4], the map Φ is Hadamard differentiable at the true copula C tangentially to a certain set $\mathbb{D}_0$ at which the distribution of $\mathbb{U}^C$ is concentrated. By (2.17), the form of the Hadamard derivative Φ'_C of Φ at C together with the functional delta method for the bootstrap [23, Theorem 3.9.11] yield conditional weak convergence in probability of $\alpha_n^\# = \sqrt{n}\{\Phi(\mathbb{G}_n^\#) - \Phi(\mathbb{C}_n^\beta)\}$ to $\mathbb{G}^C = \Phi'_C(\mathbb{U}^C)$.

Since $|\tilde{\mathbb{C}}_n - \mathbb{C}_n| \leqslant d/n$ and $|\tilde{\mathbb{C}}_n^\# - \mathbb{C}_n^\#| \leqslant d/n$ by (2.3), we obtain the conditional weak convergence in probability of $\tilde{\alpha}_n^\#$ to $\mathbb{G}^C$.

Finally, since $\alpha_n^{\beta\#}(\boldsymbol{u}) = \mathrm{E}_S[\tilde{\alpha}_n^\#(S_1/n, \ldots, S_d/n)]$ as in (2.13), we arrive at the conditional weak convergence in probability of $\alpha_n^{\beta\#}$ to $\mathbb{G}^C$ in a way similar to the proof of Proposition 2.3.1. □

References

1. Bücher A, Dette H (2010) A note on bootstrap approximations for the empirical copula process. Stat Probab Lett 80:1925–1932
2. Bücher A, Volgushev S (2013) Empirical and sequential empirical copula processes under serial dependence. J Multivar Anal 119:61–70

3. Deheuvels P (1979) La fonction de dépendence empirique et ses propriétés, un test non paramétrique d'indépendance. *Bulletin de la classe des sciences, Académie Royale de Belgique, 5e série*, 65:274–292
4. Efron B (1982) The jackknife, the bootstrap and other resampling plans. Society for Industrial and Applied Mathematics, Philadelphia
5. Fermanian J-D, Radulović D, Wegkamp MJ (2004) Weak convergence of empirical copula processes. Bernoulli 10:847–860
6. Genest C, Ghoudi K, Rivest L-P (1995) A semiparametric estimation procedure of dependence parameters in multivariate families of distributions. Biometrika 82:543–552
7. Genest C, Nešlehová J, Quessy J-F (2012) Tests of symmetry for bivariate copulas. Ann Inst Stat Math 64:811–834
8. Hall P (1992) The bootstrap and edgeworth expansion. Springer, New York
9. Hofert M, Kojadinovic I, Maechler M, Yan J, Nešlehová JG (2018) Package 'copula', R package version 0.999-19
10. Hollander M, Wolfe DA, Chicken E (2014) Nonparametric statistical methods, 3rd edn. Wiley, Hoboken
11. Khoudraji A (1995) Contributions à l'éude des copules et à la modélisation de valeurs extrêmes bivariés. PhD thesis, Université Laval, Québec, Canada
12. Kiriliouk A, Segers J, Tafakori L (2018) An estimator of the stable tail dependence function-based on the empirical beta copula. Extremes 21:581–600
13. Kojadinovic I, Yan J (2012) A non-parametric test of exchangeability for extreme-value and left-tail decreasing bivariate copulas. Scand J Stat 39:480–496
14. Kosorok MR (2008) Introduction to empirical processes and semiparametric inference. Springer, New York
15. Nelsen RB (2006) An introduction to copulas, 2nd edn. Springer, New York
16. Rémillard B, Scaillet O (2009) Testing for equality between two copulas. J Multivar Anal 100:377–386
17. Segers J (2012) Asymptotics of empirical copula processes under nonrestrictive smoothness assumptions. Bernoulli 18:764–782
18. Segers J, Sibuya M, Tsukahara H (2017) The empirical beta copula. J Multivar Anal 155:35–51
19. Shao J, Tu D (1995) The jackknife and bootstrap. Springer, New York
20. Sklar M (1959) Fonctions de répartition á n dimensions et leurs marges. Publ Inst Statist Univ Paris 8:229–231
21. Tsukahara H (2005) Semiparametric estimation in copula models. Can J Stat 33:357–375. [Erratum: Can J Stat 39:734–735 (2011)]
22. van der Vaart AW (1998) Asymptotic statistics. Cambridge University Press, Cambridge
23. van der Vaart AW, Wellner JA (1996) Weak convergence and empirical processes: with applications to statistics. Springer, New York

Chapter 3
Regression Analysis for Imbalanced Binary Data: Multi-dimensional Case

Tomonari Sei

Abstract We consider regression models for binary response data and study their behavior when the response is highly imbalanced. Previous studies have shown that if the logistic regression model is adopted, the likelihood function tends to that of an exponential family under the imbalance limit. This phenomenon is closely related to extreme value theory. In this paper, we discuss a multi-dimensional analogue of these results. First, we examine quasi-linear logistic models, where the binary outcome is explained by the log-sum-exp function of several linear scores. Then, we define a generalized model called a detectable model, and derive its imbalance limit using multivariate extreme value theory. The max-stability of the copulas corresponds to an equivariant property of the predictors.

Keywords Copula · Detectable model · Extreme value theory · Imbalanced data · Log-sum-exp function · Logistic regression · Max-stability · Quasi-linear predictor · Semi-copula

3.1 Introduction

The logistic regression model is defined by

$$P(Y = 1 \mid X = x) = G(z) = \frac{e^z}{1 + e^z}, \quad z = a + b^\top x,$$

where Y is a binary response variable, X is a p-dimensional explanatory variable, and $a \in \mathbb{R}$ and $b \in \mathbb{R}^p$ are regression coefficients. The function $G(z) = e^z/(1 + e^z)$ is the logistic distribution function, and its inverse $G^{-1}(u) = \log(u/(1 - u))$ is the logit link function.

T. Sei (✉)
The University of Tokyo, 7-3-1 Hongo, Bunkyo-ku, Tokyo 113-8656, Japan
e-mail: sei@mist.i.u-tokyo.ac.jp

N. Hoshino et al. (eds.), *Pioneering Works on Extreme Value Theory*,
JSS Research Series in Statistics,
https://doi.org/10.1007/978-981-16-0768-4_3

Now, consider the imbalanced case; that is, the probability of $Y = 1$ is very small. In the same fashion as Poisson's law of rare events, we assume that the true parameters depend on the sample size n. Specifically, let the true parameters be $a_n = -\log n + \alpha$ and $b_n = \beta$. Then, we obtain

$$\mathrm{P}(Y = 1 \mid X = x) = \frac{\frac{1}{n}e^{\alpha+\beta^\top x}}{1 + \frac{1}{n}e^{\alpha+\beta^\top x}} = \frac{1}{n}e^{\alpha+\beta^\top x} + \mathrm{O}(n^{-2}),$$

as $n \to \infty$. If the marginal distribution $F(dx)$ of X does not depend on n and its support is compact, then the weak limit of the conditional distribution of X given $Y = 1$ is, by Bayes' theorem,

$$\lim_{n\to\infty} \mathrm{P}(X \in dx \mid Y = 1) = \frac{e^{\beta^\top x} F(dx)}{\int e^{\beta^\top x} F(dx)}, \tag{3.1}$$

which is an exponential family [13]. Furthermore, the joint distribution of X and Y converges to an inhomogeneous Poisson point process with intensity measure $e^{\alpha+\beta^\top x} F(dx)$; see [17] for details. We call the limit of a regression model under the imbalance assumption the *imbalance limit*.

There are other binary regression models with the same imbalance limit. For example, the complementary log-log link, which corresponds to $G(z) = 1 - \exp(-e^z)$, has the same imbalance limit as the logit link. In this case, $G(z)$ is the negative Gumbel distribution function, one of the min-stable distributions.

Similarly, the limit of a binary regression model with a cumulative distribution function $G(z)$ is characterized by extreme value theory [15]. Models with distinct link functions have the same imbalance limit if the corresponding distribution functions belong to the same domain of attraction. Here, min-stability corresponds to stability with respect to a resolution change of the explanatory variables [2].

In this study, we develop a multivariate analogue of the above facts. The function G is generalized to include multi-dimensional functions. A practical class is the quasi-linear logistic regression model proposed by [12], which combines several linear predictors using the log-sum-exp function. See Sect. 3.2 for a precise definition. We define a generalized class, called a detectable model. The imbalance limit of the model is obtained using the multivariate extreme value theory (e.g., [4, 14, 16]). Here, the max-stability of the copulas corresponds to an equivariant property of the detectable predictors.

The rest of the paper is organized as follows. In Sect. 3.2, we review the quasi-linear logistic regression model. The model is further generalized in Sect. 3.3, and the imbalance limit is studied in Sect. 3.4. Examples of equivariant predictors are provided in Sect. 3.5. Finally, Sect. 3.6 concludes the paper.

3.2 Quasi-linear Logistic Regression Model and Its Imbalance Limit

In this section, we first define the quasi-linear logistic regression model, and then derive its imbalance limit, as in (3.1).

3.2.1 The Quasi-linear Logistic Regression Model

Omae et al. [12] define a *quasi-linear logistic regression model* as follows:

$$\mathrm{P}(Y = 1 \mid X = x) = \frac{e^{Q}}{1 + e^{Q}}, \tag{3.2}$$

$$Q = \frac{1}{\tau} \log \left(\sum_{k=1}^{K} e^{\tau(a_k + b_k^\top x)} \right), \tag{3.3}$$

where X is a p-dimensional explanatory variable, $a_k \in \mathbb{R}$ and $b_k \in \mathbb{R}^p$ are regression coefficients for each $k = 1, \ldots, K$, and $\tau > 0$ is a tuning parameter. It is also possible to define (3.3) for $\tau < 0$ (see [11]), but we restrict τ to be positive, owing to a property discussed later (Lemma 3.1 in Sect. 3.3). We assume $K \geq 2$, unless otherwise stated.

The model reduces to the logistic regression model if $K = 1$, but is not even identifiable with respect to the regression coefficients if $K \geq 2$. Therefore, some restrictions and regularizations are imposed in practice. For example, the explanatory variable X is partitioned into K subvectors $X_{(1)}, \ldots, X_{(K)}$ using a clustering method such as the K-means method. Then, the coordinates of b_k, except for those corresponding to $X_{(k)}$, are set to zero for each k.

Denote the K linear predictors by $z_k = a_k + b_k^\top x$. Then, the right-hand side of (3.3) is written as

$$Q(z_1, \ldots, z_K) = \frac{1}{\tau} \log \left(\sum_k e^{\tau z_k} \right),$$

which we call the *quasi-linear predictor* or the *log-sum-exp function* (refer to [3]). The log-sum-exp function tends to the simple sum $\sum_k z_k$ as $\tau \to 0$ up to a constant term, and tends to $\max(z_1, \ldots, z_K)$ as $\tau \to \infty$ for fixed $(z_1, \ldots, z_K)$.

Reference [12] proposed the following generalized class:

$$Q = \phi^{-1} \left(\sum_k \phi(z_k) \right), \tag{3.4}$$

where ϕ is an invertible function. The log-sum-exp function is a particular case of $\phi(z) = e^{\tau z}$. Further generalization is discussed in the next section. In what follows, we call (3.4) the *generalized quasi-linear predictor* with the *generator* ϕ.

Remark 3.1 In [11], a slightly different definition is used,

$$Q = \phi^{-1}\left(\frac{1}{K}\sum_k \phi(z_k)\right),$$

and is called the generalized average or the Kolmogorov–Nagumo average. The difference is the factor $1/K$. In this study, we adopt the form in (3.4) because we focus on a property shown in Lemma 3.1, later.

3.2.2 Imbalance Limit

We derive the imbalance limit of the quasi-linear logistic regression model. Suppose the true parameters a_k and b_k in (3.3) are given by

$$a_{k,n} = -\log n + \alpha_k, \quad b_{k,n} = \beta_k,$$

which depend on the sample size n. Then, we have

$$Q = -\log n + \frac{1}{\tau}\log\left(\sum_{k=1}^{K} e^{\tau(\alpha_k + \beta_k^\top x)}\right),$$

and obtain the asymptotic form

$$\mathrm{P}(Y = 1 \mid X = x) = \frac{e^Q}{1 + e^Q} = \frac{1}{n}\left(\sum_k e^{\tau(\alpha_k + \beta_k^\top x)}\right)^{1/\tau} + \mathrm{O}(n^{-2}).$$

The conditional distribution of X, given $Y = 1$, is

$$\begin{aligned}\mathrm{P}(X \in dx \mid Y = 1) &= \frac{\mathrm{P}(Y = 1 \mid X = x)F(dx)}{\int \mathrm{P}(Y = 1 \mid X = x)F(dx)} \quad \text{(Bayes' theorem)} \\ &\to \frac{\{\sum_k e^{\tau(\alpha_k + \beta_k^\top x)}\}^{1/\tau} F(dx)}{\int \{\sum_k e^{\tau(\alpha_k + \beta_k^\top x)}\}^{1/\tau} F(dx)},\end{aligned}$$

where $F(dx)$ is the marginal distribution of X. In particular, the distribution is reduced to a mixed exponential family if $\tau = 1$.

Remark 3.2 In [12], the authors note that the quasi-linear logistic model with $\tau = 1$ is Bayes optimal if the conditional distribution of X, given Y, is mixture normal.

Specifically, suppose that the ratio of the conditional distributions of X is a mixture exponential family

$$\frac{P(X \in dx \mid Y = 1)}{P(X \in dx \mid Y = 0)} = \frac{1}{Z} \sum_k e^{\alpha_k + \beta_k^\top x},$$

where α_k and β_k are parameters, and Z is a normalization constant. Then, the logit of the predictive distribution is

$$\log \frac{P(Y = 1 \mid X = x)}{P(Y = 0 \mid X = x)} = \log \left(\sum_k e^{\alpha_k^* + \beta_k^\top x} \right),$$

where $\alpha_k^* = \alpha_k - \log Z + \log(\pi_1/\pi_0)$ and $\pi_y = P(Y = y)$. This is the quasi-linear predictor.

3.3 Extension of the Model and Its Copula Representation

In this section, we extract several features of the quasi-linear logistic model, and use these to define a generalized class of regression models. We also discuss the relationship between this class of models and copula theory.

3.3.1 Detectable Model

We first focus on the following property of the generalized quasi-linear predictor (3.4), with generator ϕ.

Lemma 3.1 *Suppose $\phi : \mathbb{R} \to (0, \infty)$ is continuous, strictly increasing, and has boundary values $\phi(-\infty) = 0$ and $\phi(\infty) = \infty$. Then, the generalized quasi-linear predictor (3.4) satisfies*

$$Q(z_1, \ldots, z_k, \ldots, z_K) \text{ is increasing in } z_k, \tag{3.5}$$

$$Q(-\infty, \ldots, z_k, \ldots, -\infty) = z_k, \tag{3.6}$$

for each k and $(z_1, \ldots, z_K) \in \mathbb{R}^K$.

Properties (3.5) and (3.6) are also satisfied by $Q(z_1, \ldots, z_K) = \max(z_1, \ldots, z_K)$, where the increasing property in (3.5) is interpreted as nondecreasing. In a sense, property (3.6) respects the maximum of the K linear scores (p. 4 of [12]).

In general, we call a function $Q : \mathbb{R}^K \to \mathbb{R}$ a *detectable predictor* if it satisfies (3.5) and (3.6). Note that the quantity $Q(-\infty, \ldots, z_k, \ldots, -\infty)$ does not depend on the choice of the diverging sequence of $(z_1, \ldots, z_{k-1}, z_{k+1}, \ldots, z_K)$ to $(-\infty, \ldots, -\infty)$ under the monotonicity condition (3.5).

Remark 3.3 The term "detectable" is borrowed from the neural network literature (e.g., Chaps. 6 and 9 of [8]), where a number of compositions of one-dimensional nonlinear functions and multi-dimensional linear functions are applied. In contrast, we focus on the properties of the multi-dimensional nonlinear function Q using the copula theory.

Then, we define a model class, as follows.

Definition 3.1 (*Detectable model*) Let Q be a detectable predictor, and let G_1 be a strictly increasing continuous distribution function. Then, a *detectable model* with G_1 and Q is defined by

$$\mathrm{P}(Y = 1 \mid X = x) = G_1(Q), \tag{3.7}$$

$$Q = Q(a_1 + b_1^\top x, \ldots, a_K + b_K^\top x). \tag{3.8}$$

We call G_1 the *inverse link function.*

For example, the quasi-linear logistic model is a detectable model with $G_1(Q) = e^Q/(1 + e^Q)$ and $Q(z_1, \ldots, z_K) = \tau^{-1} \log(\sum_k e^{\tau z_k})$. Similarly to the quasi-linear model, the detectable model aggregates K linear predictors into a quantity Q.

We give two properties of detectable predictors. The proofs are easy, and thus are omitted.

Lemma 3.2 *Any detectable predictor Q satisfies an inequality*

$$Q(z_1, \ldots, z_K) \geq \max(z_1, \ldots, z_K).$$

Lemma 3.3 *Let Q_1 and Q_2 be detectable predictors. Then,* $(Q_1 + Q_2)/2$, $\max(Q_1, Q_2)$ *and* $\min(Q_1, Q_2)$ *are also detectable. More generally, if a function $f : \mathbb{R}^2 \to \mathbb{R}$ is increasing in each argument and satisfies $f(x, x) = x$, for all $x \in \mathbb{R}$, then $f(Q_1, Q_2)$ is detectable.*

The generalized average mentioned in Remark 3.1 is an example of f in which $f(x, x) = x$.

3.3.2 Copula Representation

The detectable model has a copula representation. Consider a detectable model with an inverse link function G_1, and a detectable predictor Q. Denote the composite map of G_1 and Q by

$$G(z_1, \ldots, z_K) = G_1(Q(z_1, \ldots, z_K)).$$

Then, G is increasing in each variable and satisfies

$$G(-\infty, \ldots, z_k, \ldots, -\infty) = G_1(z_k).$$

Next, define a dual of G by

$$H(w_1, \ldots, w_K) = 1 - G(-w_1, \ldots, -w_K), \quad (w_1, \ldots, w_K) \in \mathbb{R}^K,$$

and

$$H_1(w) = 1 - G_1(-w), \quad w \in \mathbb{R}. \tag{3.9}$$

Then, H is increasing in each variable and satisfies

$$H(\infty, \ldots, w_k, \ldots, \infty) = H_1(w_k).$$

Thus, H_1 is considered the kth marginal distribution function of H. Note that H itself may not be a multivariate distribution function because the K-increasing property may fail. Recall that a function H is said to be *K-increasing* if $\Delta_1 \cdots \Delta_K H \geq 0$, where Δ_k is the difference operator with respect to the kth argument.

Finally, as with Sklar's theorem, we define

$$C(u_1, \ldots, u_K) = H(H_1^{-1}(u_1), \ldots, H_1^{-1}(u_K)). \tag{3.10}$$

Then, C satisfies the following conditions:

$$C(1, \ldots, u_k, \ldots, 1) = u_k,$$
$$C(u_1, \ldots, u_K) \text{ is increasing in } u_k,$$

for each k. A function $C : [0, 1]^K \to [0, 1]$ satisfying the two conditions is called a *semi-copula* (see Chap. 8 of [5]). Any copula is a semi-copula, but the converse is not true. The Kth-order difference $\Delta_1 \cdots \Delta_K C$ of a semi-copula, which measures a rectangular region, may be negative.

We summarize this result as follows.

Theorem 3.1 (Copula representation) *A detectable model specified by an inverse link function G_1 and a detectable predictor Q is represented as*

$$\begin{aligned} G_1(Q(z_1, \ldots, z_K)) &= G(z_1, \ldots, z_K) \\ &= 1 - H(-z_1, \ldots, -z_K) \\ &= 1 - C(H_1(-z_1), \ldots, H_1(-z_K)), \end{aligned}$$

where C *is a semi-copula, and* H_1 *is a univariate continuous distribution function. The correspondence*

$$\{G_1, Q\} \leftrightarrow \{H_1, C\}$$

is one-to-one.

Proof It is sufficient to prove the one-to-one correspondence. Indeed, if G_1 and Q are given, then H_1 and C are determined by (3.9) and (3.10), respectively. Conversely, if H_1 and C are given, then we have $G_1(z) = 1 - H_1(-z)$ by (3.9), and

$$Q(z_1, \ldots, z_K) = -H_1^{-1}(C(H_1(-z_1), \ldots, H_1(-z_K)))$$

holds. □

Consider again the quasi-linear logistic regression model with the log-sum-exp predictor, which corresponds to (3.2) and (3.3). Then, the functions H_1 and C in Theorem 3.1 are the logistic distribution function and

$$C(u_1, \ldots, u_K) = \frac{1}{1 + (\sum_{k=1}^{K} (\frac{1-u_k}{u_k})^\tau)^{1/\tau}}, \tag{3.11}$$

respectively. The function C is a copula if $\tau \geq 1$, as shown in Example 4.26 of [10]. In particular, if $\tau = 1$, then

$$C(u_1, \ldots, u_K) = \frac{1}{1 + \sum_{k=1}^{K} \frac{1-u_k}{u_k}},$$

which belongs to the Clayton copula family [10]. If $\tau \to \infty$, then C converges to $\min_k u_k$, the upper Fréchet–Hoeffding bound. If $0 < \tau < 1$, C is not a copula, in the strict sense, because it is not K-increasing.

We say that a semi-copula C is Archimedean if it is written as

$$C(u_1, \ldots, u_K) = \psi^{-1}(\psi(u_1) + \cdots + \psi(u_K)),$$

with a decreasing function $\psi : (0, 1) \to (0, \infty)$ called the generator (e.g., [10]). For example, the semi-copula in (3.11) is Archimedean with the generator $\psi(u) = (\frac{1-u}{u})^\tau$.

Archimedean semi-copulas characterize the generalized quasi-linear models as stated in the following theorem. The proof is straightforward.

Theorem 3.2 (Archimedean case) *Let* $\{G_1, Q\}$ *be a detectable model and* $\{H_1, C\}$ *be the corresponding pair determined by Theorem 3.1. Then,* Q *is a generalized quasi-linear predictor (3.4) with a generator* ϕ *if and only if* C *is an Archimedean semi-copula with a generator* ψ*. The relation between the generators* ϕ *and* ψ *is given by* $\phi(z) = \psi(H_1(-z))$.

Note that ψ depends not just on ϕ, but also on the inverse link G_1.

Remark 3.4 Here, we briefly discuss the merit of having a genuine copula in the copula representation of a detectable model, where a genuine copula means a semi-copula with the K-increasing property. If C is a genuine copula, then the detectable model $P(Y = 1 \mid X = x) = G_1(Q(z_1, \ldots, z_K))$ has the following latent variable representation. Take a random vector $U = (U_1, \ldots, U_K)$, distributed according to the copula C. Then, $G_1(Q(z_1, \ldots, z_K)) = 1 - C(H_1(-z_1), \ldots, H_1(-z_K))$ coincides with the probability of an event, such that $U_k > H_1(-z_k)$ for at least one k. Now, the response variable Y can be assumed to be the indicator function of that event. The random vector U is seen as a latent variable. Once a latent variable representation is obtained, we can also consider a state-space model for time-dependent data. This is left to future research.

3.4 The Imbalance Limit of Detectable Models

In this section, we characterize the imbalance limit of detectable models using the copula representation in Theorem 3.1 and the multivariate extreme value theory [4, 14, 16].

Recall that detectable models are specified by a univariate distribution function H_1 and a semi-copula C. Throughout this section, we fix H_1 as the Gumbel distribution function

$$H_1(w) = \exp(-e^{-w}),$$

and focus on the semi-copulas C. In this case, the inverse link function is $G_1(z) = 1 - \exp(-e^{z})$, which corresponds to the complementary log-log link function. The relation between C and the detectable predictor Q is given by

$$C(u_1, \ldots, u_K) = \exp(-e^{Q(z_1, \ldots, z_K)}), \quad u_k = \exp(-e^{z_k}), \tag{3.12}$$

by Theorem 3.1.

A semi-copula C is said to be *extreme* if there exists a semi-copula C_0 such that

$$C(u_1, \ldots, u_K) = \lim_{n\to\infty} C_0^n(u_1^{1/n}, \ldots, u_K^{1/n}), \quad u \in [0, 1]^K. \tag{3.13}$$

In this case, we say that C_0 belongs to the domain of attraction of C. A semi-copula C is said to be *max-stable* if, for all $n \geq 1$,

$$C(u_1, \ldots, u_K) = C^n(u_1^{1/n}, \ldots, u_K^{1/n}), \quad u \in [0, 1]^K. \tag{3.14}$$

The following lemma is widely known for copulas.

Lemma 3.4 *A semi-copula C is extreme if and only if it is max-stable.*

Proof It is obvious that any max-stable semi-copula is also extreme. Conversely, assume that C is extreme. Let C_0 be a semi-copula that satisfies (3.13). Then, we have

$$\begin{aligned} C^m(u_1^{1/m}, \ldots, u_K^{1/m}) &= \lim_{n\to\infty} C_0^{nm}(u_1^{1/nm}, \ldots, u_K^{1/nm}) \\ &= C(u_1, \ldots, u_K), \end{aligned}$$

for all $m \geq 1$, which means C is max-stable. □

Max-stability is reflected in detectable models as follows.

Lemma 3.5 *Consider a detectable model specified by the Gumbel distribution function H_1 and a semi-copula C. Then, C is max-stable if and only if the detectable predictor Q is equivariant with respect to location; that is,*

$$Q(z_1 + \alpha, \ldots, z_K + \alpha) = Q(z_1, \ldots, z_K) + \alpha, \quad \alpha \in \mathbb{R}. \tag{3.15}$$

Proof Let C be max-stable. Then, by (3.12) and (3.14), we have

$$\begin{aligned} Q(z_1, \ldots, z_K) &= \log(-\log C(\exp(-e^{z_1}), \ldots, \exp(-e^{z_K}))) \\ &= \log\left(-\log C^n\left(\exp\left(-\frac{1}{n}e^{z_1}\right), \ldots, \exp\left(-\frac{1}{n}e^{z_K}\right)\right)\right) \\ &= \log n + Q(-\log n + z_1, \ldots, -\log n + z_K), \end{aligned}$$

for all $n \geq 1$. Then, (3.15) is proved for $\alpha = \log x$, with positive rational numbers x. The result follows from the monotonicity of Q. The converse is proved in a similar manner. □

Remark 3.5 According to extreme value theory, the stable tail dependence function corresponding to a max-stable copula C is defined by

$$l(x_1, \ldots, x_K) = -\log C(e^{-x_1}, \ldots, e^{-x_K}), \quad (x_1, \ldots, x_K) \in [0, \infty)^K,$$

which satisfies a homogeneous property $l(tx_1, \ldots, tx_K) = tl(x_1, \ldots, x_K)$ (see [6]). The equivariance of Q in Lemma 3.5 is interpreted as another representation of the max-stable property. Note that l is not suitable for constructing predictors because its domain is not the whole space.

The imbalance limit of detectable models is characterized as follows. The result is an analogue of that in extreme value theory (e.g., Corollary 6.1.3 of [4]).

Theorem 3.3 (Imbalance limit) *Consider a detectable model specified by the Gumbel distribution function H_1 and a semi-copula C. Let G_1, Q, and G be the functions determined by Theorem 3.1. Then, the following three conditions are equivalent, where $\bar{Q}$ denotes an equivariant predictor:*

1. *The predictor Q admits a limit*

$$\lim_{n\to\infty}\{Q(z_1-\log n,\ldots,z_K-\log n)+\log n\}=\bar{Q}(z_1,\ldots,z_K).$$

2. *The function G admits a limit*

$$\lim_{n\to\infty}\{n\,G(z_1-\log n,\ldots,z_K-\log n)\}=e^{\bar{Q}(z_1,\ldots,z_K)}.$$

3. *The semi-copula C belongs to the domain of attraction of*

$$\bar{C}(u_1,\ldots,u_K)=\exp(-e^{\bar{Q}(z_1,\ldots,z_K)}),\quad u_k=\exp(-e^{z_k}).$$

Under these conditions, if the true regression coefficients are $a_{k,n}=-\log n+\alpha_k$ *and* $b_{k,n}=\beta_k$*, then the weak limit of the conditional distribution of X is*

$$\lim_{n\to\infty}P(X\in dx\mid Y=1)=\frac{e^{\bar{Q}(\alpha_1+\beta_1^\top x,\ldots,\alpha_K+\beta_K^\top x)}F(dx)}{\int e^{\bar{Q}(\alpha_1+\beta_1^\top x,\ldots,\alpha_K+\beta_K^\top x)}F(dx)}$$

whenever the support of $F(dx)=P(X\in dx)$ *is compact.*

Proof The equivalence of conditions 1 and 3 follows immediately from the relation (3.12). We prove the equivalence of conditions 1 and 2. Because

$$G(z_1,\ldots,z_K)=1-\exp(-e^{Q(z_1,\ldots,z_K)}),$$

condition 2 is written as

$$\lim_{n\to\infty}n\{1-\exp(-e^{Q(z_1-\log n,\ldots,z_K-\log n)})\}=e^{\bar{Q}(z_1,\ldots,z_K)},$$

which is also equivalent to

$$\lim_{n\to\infty}ne^{Q(z_1-\log n,\ldots,z_K-\log n)}=e^{\bar{Q}(z_1,\ldots,z_K)}.$$

The logarithm of both sides yields condition 1.

Next, we show the convergence of the conditional distribution. Note that the convergence of G in condition 2 is locally uniform with respect to $(z_1,\ldots,z_K)$ because G is monotone in each argument. Then, Bayes' theorem and the compactness of the support of F imply

$$\begin{aligned}P(X\in dx\mid Y=1)&=\frac{G(-\log n+\alpha_1+\beta_1^\top x,\ldots,-\log n+\alpha_K+\beta_K^\top x)F(dx)}{\int G(-\log n+\alpha_1+\beta_1^\top x,\ldots,-\log n+\alpha_K+\beta_K^\top x)F(dx)}\\&\to\frac{e^{\bar{Q}(\alpha_1+\beta_1^\top x,\ldots,\alpha_K+\beta_K^\top x)}F(dx)}{\int e^{\bar{Q}(\alpha_1+\beta_1^\top x,\ldots,\alpha_K+\beta_K^\top x)}F(dx)},\end{aligned}$$

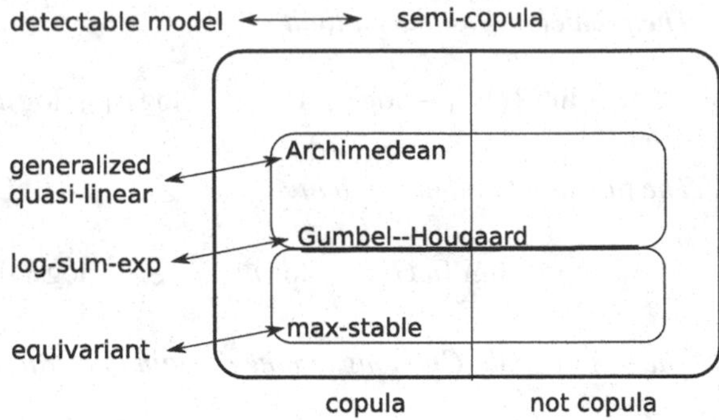

Fig. 3.1 Classification of detectable models, where H_1 is fixed to be Gumbel

as stated. □

For example, consider the semi-copula in (3.11), which is derived from the log-sum-exp logistic model. Here, the extreme semi-copula $\bar{C}$ in Theorem 3.3 is

$$\begin{aligned}\bar{C}(u_1, \ldots, u_K) &= \lim_{n\to\infty} C^n(u_1^{1/n}, \ldots, u_K^{1/n}) \\ &= \lim_{n\to\infty} \left(\frac{1}{1 + (\sum_{k=1}^K (u_k^{-1/n} - 1)^\tau)^{1/\tau}} \right)^n \\ &= \exp\left(- \left(\sum_{k=1}^K (-\log u_k)^\tau \right)^{1/\tau} \right),\end{aligned}$$

which is called the Gumbel–Hougaard copula [10] if $\tau \geq 1$. In particular, it reduces to the independent copula if $\tau = 1$. The Gumbel–Hougaard copula is an Archimedean copula with generator $\psi(u) = (-\log u)^\tau$. In fact, this class is characterized by the max-stable Archimedean property [7].

The detectable predictor Q corresponding to the Gumbel–Hougaard copula when H_1 is Gumbel is the log-sum-exp

$$\begin{aligned}Q(z_1, \ldots, z_K) &= \log(-\log \bar{C}(\exp(-e^{z_1}), \ldots, \exp(-e^{z_K}))) \\ &= \frac{1}{\tau} \log \left(\sum_{k=1}^K e^{\tau z_k} \right).\end{aligned}$$

As a result, the generalized quasi-linear predictor with the equivariant property (3.15) is limited to be the log-sum-exp predictor. This fact is directly confirmed in Lemma 3.6 in Sect. 3.5. Note too that the independent copula corresponds to $\tau = 1$.

Figure 3.1 classifies the detectable models.

If H_1 is not Gumbel, the imbalance limit depends on the domain of attraction to which H_1 belongs. For example, the logistic distribution belongs to the domain of attraction of the Gumbel. For such a case, the statements in Theorem 3.3 still hold.

3.5 Examples of Equivariant Predictors

In this section, we provide examples of equivariant predictors, where the equivariance is defined by (3.15). Recall that equivariant predictors correspond to max-stable semi-copulas if H_1 is Gumbel (Lemma 3.5). In the following, we construct the predictors directly and do not use the copula representations (except for Lemma 3.7).

It is obvious that, by definition, the log-sum-exp predictor is equivariant. Conversely, the log-sum-exp predictor is characterized as follows.

Lemma 3.6 *Let Q be a generalized quasi-linear predictor with a generator ϕ, where $\phi : \mathbb{R} \to (0, \infty)$ is continuous and strictly increasing, $\phi(-\infty) = 0$, and $\phi(\infty) = \infty$. Then, Q is equivariant if and only if it is the log-sum-exp predictor for some $\tau > 0$.*

Proof We prove the "only if" part. It is enough to consider the case $K = 2$, because we can set $\phi(z_k) = 0$ for $3 \le k \le K$ by letting $z_k \to -\infty$. Because Q is equivariant, we have

$$\phi^{-1}(\phi(z_1 + \alpha) + \phi(z_2 + \alpha)) = \phi^{-1}(\phi(z_1) + \phi(z_2)) + \alpha,$$

for any $\alpha \in \mathbb{R}$. Applying ϕ to the both sides and putting $z_k = \phi^{-1}(x_k)$, we obtain

$$\phi(\phi^{-1}(x_1) + \alpha) + \phi(\phi^{-1}(x_2) + \alpha) = \phi(\phi^{-1}(x_1 + x_2) + \alpha).$$

This is Cauchy's functional equation (Theorem 2.1.1 of [1]) on $\eta(x) := \phi(\phi^{-1}(x) + \alpha)$. Because η is increasing, the solution has to be $\eta(x) = \phi(\phi^{-1}(x) + \alpha) = \sigma_\alpha x$, for some $\sigma_\alpha > 0$. Put $z = \phi^{-1}(x)$ to obtain $\phi(z + \alpha) = \sigma_\alpha \phi(z)$. By letting $z = 0$, we have $\sigma_\alpha = \phi(\alpha)/\phi(0)$ and, therefore,

$$\phi(z + \alpha) = \frac{\phi(\alpha)\phi(z)}{\phi(0)}.$$

By putting $\psi(z) = \log \phi(z) - \log \phi(0)$, we have $\psi(z + \alpha) = \psi(z) + \psi(\alpha)$. Again, because ψ is increasing, we have $\psi(z) = \tau z$, for some $\tau > 0$, which means $\phi(z) = \phi(0)e^{\tau z}$. Hence, ϕ is the generator of the log-sum-exp predictor. □

For other examples, consider

$$Q(z_1, z_2) = \frac{z_1 + z_2 + \sqrt{(z_1 - z_2)^2 + 4\varepsilon^2}}{2}, \tag{3.16}$$

where $\varepsilon > 0$ is a fixed constant. This is actually an equivariant detectable predictor. Indeed, it satisfies the conditions $\partial Q/\partial z_k > 0$, $Q(z, -\infty) = Q(-\infty, z) = z$, and $Q(z_1 + \alpha, z_2 + \alpha) = Q(z_1, z_2) + \alpha$. The function Q in (3.16) is quite different from the log-sum-exp function when $|z_1 - z_2|$ is large. Indeed, if $z_1 > z_2$ and z_1 is fixed, then

$$Q(z_1, z_2) = z_1 + \mathrm{O}((z_1 - z_2)^{-1}),$$

as $z_1 - z_2 \to \infty$, whereas

$$\frac{1}{\tau}\log(e^{\tau z_1} + e^{\tau z_2}) = z_1 + \mathrm{O}(e^{-\tau(z_1 - z_2)}).$$

The case of $z_1 < z_2$ is derived in a similar manner. The behavior for large $|z_1 - z_2|$ may affect the numerical stability of the parameter estimation. This is left to future work.

A multivariate extension of (3.16) is the unique solution of

$$\prod_{k=1}^{K}(Q - z_k) = \varepsilon^K, \quad Q > \max(z_1, \ldots, z_K), \tag{3.17}$$

which we call an *algebraic predictor*. The tail behavior is given by

$$Q = z_{(1)} + \mathrm{O}((z_{(1)} - z_{(2)})^{-(K-1)}),$$

as $z_{(1)} - z_{(2)} \to \infty$, where $z_{(1)} \geq \cdots \geq z_{(K)}$ is the order statistic of $z_1, \ldots, z_K$.

We can construct a broad class of equivariant predictors using a direct consequence of extreme value theory, as follows.

Lemma 3.7 *Let μ be a (nonnegative) measure on the simplex $\Delta = \{s \mid \sum_{k=1}^{K} s_k = 1, s_1, \ldots, s_K \geq 0\}$, such that $\int s_k \mu(ds) = 1$ for all k. Then,*

$$Q(z_1, \ldots, z_K) = \log \int \max(s_1 e^{z_1}, \ldots, s_K e^{z_K})\mu(ds) \tag{3.18}$$

is an equivariant detectable predictor. Conversely, if Q is equivariant and the semi-copula C determined by Theorem 3.1 with the Gumbel marginal H_1 is a (genuine) copula, then there exists such a unique measure μ.

Proof It is easy to see that Q in (3.18) is actually an equivariant predictor. To prove the converse, suppose that Q is equivariant and C determined by Theorem 3.1, with the Gumbel marginal H_1, is a copula. Lemma 3.5 implies that C is a max-stable copula and, therefore, H in Theorem 3.1 is a max-stable distribution function with the Gumbel marginal H_1. Then, by Proposition 5.11′ of [14], H has the spectral representation

$$H(x_1, \ldots, x_K) = \exp\left\{-\int_\Delta \max(s_1 e^{-x_1}, \ldots, s_K e^{-x_K})\mu(dx)\right\},$$

with a measure μ on Δ such that $\int s_k \mu(dx) = 1$, for all k. Equation (3.18) follows from the representation $Q(z_1, \ldots, z_K) = \log(-\log H(-z_1, \ldots, -z_K))$. □

The measure μ is called the *spectral measure*. For example, let $K = 3$ and $\mu = (\delta_{(1/2,1/2,0)} + \delta_{(1/2,0,1/2)} + \delta_{(0,1/2,1/2)})/2$, where δ denotes the Dirac measure. Then,

$$Q(z_1, z_2, z_3) = \log\left(\frac{e^{\max(z_1,z_2)} + e^{\max(z_1,z_3)} + e^{\max(z_2,z_3)}}{2}\right).$$

Using the order statistic $z_{(1)} \geq z_{(2)} \geq z_{(3)}$ of (z_1, z_2, z_3), we have

$$Q(z_1, z_2, z_3) = \log\left(e^{z_{(1)}} + \frac{e^{z_{(2)}}}{2}\right),$$

which, in particular, depends only on the top two scores $(z_{(1)}, z_{(2)})$. More generally,

$$Q(z_1, \ldots, z_K) = \frac{1}{\tau}\log\left(e^{\tau z_{(1)}} + \sum_{k=2}^{K} \lambda_k e^{\tau z_{(k)}}\right)$$

is equivariant for any positive τ and nonnegative λ_k. The log-sum-exp function is the special case $\lambda_2 = \cdots = \lambda_K = 1$.

Note that Q defined by (3.18) must satisfy

$$Q(z_1, \ldots, z_K) \leq \log(e^{z_1} + \cdots + e^{z_K}),$$

which follows from $\max_k(s_k e^{z_k}) \leq \sum_k s_k e^{z_k}$. In particular, employing the lower bound in Lemma 3.2, we can prove that the tail behavior of Q is restricted to

$$Q(z_1, \ldots, z_K) = z_{(1)} + O(e^{-(z_{(1)} - z_{(2)})}),$$

as $z_{(1)} - z_{(2)} \to \infty$. Thus, the algebraic predictor (3.17) cannot be expressed as (3.18) with a (nonnegative) spectral measure μ.

3.6 Conclusion

In this paper, we introduced detectable models as generalizations of the quasi-linear logistic models, and then derived the imbalance limit (Theorem 3.3). A key property is that of equivariance (3.15). The log-sum-exp function is characterized as a unique equivariant quasi-linear predictor (Lemma 3.6); see Sect. 3.5 for examples of other equivariant predictors.

We have not conducted any simulation results. Thus, future work should investigate the numerical stability of the maximum likelihood estimator when an equivariant predictor such as the algebraic predictor (3.17) is adopted.

The generalized average of the form in Remark 3.1 can be extended to functions with the property $Q(z, \ldots, z) = z$ instead of (3.6). Regression models with such a property may exhibit different behaviors.

Lastly, we have implicitly assumed that the conditional probability $P(Y = 1 \mid X = x)$ ranges from zero to one. However, this assumption may be relaxed. In fact,

[9] suggests an asymmetric logistic regression model that uses $G(z) = (e^z + \kappa)/(1 + e^z + \kappa)$, for $\kappa > 0$, as the inverse link function. This function is not even a distribution function because $G(-\infty) > 0$. Therefore, it would be interesting to investigate what happens if $\kappa_n \to 0$ as $n \to \infty$ under the imbalance limit.

Acknowledgements The author thanks the reviewer for his/her insightful comments and references. In particular, the author was not previously aware of the term "semi-copula." This research was motivated by Prof. Masaaki Sibuya's questions during a presentation at Keio University in 2013, and his subsequent comments during the workshop at the Institute of Statistical Mathematics in 2019. The author thanks Katsuhiro Omae, Osamu Komori, and Shinto Eguchi for providing helpful discussions and information. This work was supported by JSPS KAKENHI Grant Numbers 26108003 and 17K00044.

References

1. Aczél J (1966) Lectures on functional equations and their applications. Academic, New York
2. Baddeley A, Berman M, Fisher NI, Hardegen A, Milne RK, Schuhmacher D, Shah R, Turner R (2010) Spatial logistic regression and change-of-support in Poisson point processes. Electron J Statist 4:1151–1201
3. Boyd S, Vandenberghe L (2004) Convex optimization. Cambridge University Press, Cambridge
4. de Haan L, Ferreira A (2006) Extreme value theory – an introduction. Springer, Berlin
5. Durante F, Sempi C (2016) Principles of copula theory. CRC Press, Boca Raton
6. Genest C, Nešlehová J (2012) Copula modeling for extremes. In: El-Shaarawi AH, Piegorsch WW (eds) Encyclopedia of environmetrics, 2nd ed. Wiley, Hoboken
7. Genest C, Rivest L-P (1989) A characterization of Gumbel's family of extreme value distributions. Stat Probab Lett 8:207–211
8. Goodfellow I, Bengio Y, Courville A (2016) Deep learning. The MIT Press, Cambridge
9. Komori O, Eguchi S, Ikeda S, Okamura H, Ichinokawa M, Nakayama S (2016) An asymmetric logistic regression model for ecological data. Methods Ecol Evol 7:249–260
10. Nelsen RB (1999) An introduction to copula, 2nd edn. Springer, Berlin
11. Omae K (2017) Statistical learning by quasi-linear predictor. SOKENDAI, Ph. D. Thesis
12. Omae K, Komori O, Eguchi S (2017) Quasi-linear score for capturing heterogeneous structure in biomarkers. BMC Bioinform 18(308):1–15
13. Owen AB (2007) Infinitely imbalanced logistic regression. J Mach Learn Res 8:761–773
14. Resnick SI (1987) Extreme values, regular variation, and point processes. Springer, New York
15. Sei T (2014) Infinitely imbalanced binomial regression and deformed exponential families. J Stat Plann Inference 149:116–124
16. Sibuya M (1960) Bivariate extreme statistics. I Ann Inst Stat Math 11(3):195–210
17. Warton DI, Shepherd LC (2010) Poisson point process models solve the "pseudo-absence problem" for presence only data in ecology. Ann Appl Stat 4:1383–1402

Chapter 4
An Analysis of Extremes: Semiparametric Efficiency in Regression

Akichika Ozeki and Kjell Doksum

Abstract Let Y denote a response variable and X denote a covariate. We consider statistical inferences in semiparametric regression models, where the parameter of interest is the upper boundary $\theta(x) = g(x\beta)$ of the support $[0, \theta(x)]$ of the conditional distribution of Y, given $X = x$, where the boundary structure $g(\cdot)$ is known. By extending Le Cam's theory of limits of experiments to include semiparametric models, we construct estimators of the boundary parameter $\theta(x)$, with bias correction. The risk of each estimator reaches a lower bound for the one-sample problem, two-sample problem, and one continuous covariate (predictor) case. When $g(\cdot)$ is unknown with one continuous predictor X, we propose a selection method for the boundary structure $g(\cdot)$ under regularity conditions. This yields consistent estimators of $g(\cdot)$ and β, although efficiency is not proved under this general model. We conclude with a real-data study on the longevity of lung cancer patients.

Keywords Adaptive estimators · Boundary parameters · Continuous predictor · Extreme values · Limits of experiments · Semiparametric efficiency

4.1 Introduction

Extremes are the smallest and largest possible values of a response variable Y and, in many cases, may depend on a covariate X. Such extremes are of interest in studies on earthquakes, radiation, floods, wind, breaking strength, rare events, and longevity, among others. In this study, extremes are modeled as the upper boundary $\theta(x)$ of the support $[0, \theta(x)]$ of the conditional distribution of a response Y, given $X = x$,

---

A. Ozeki (✉)
Eli Lilly Japan K.K., 5-1-28, Isogami-Dori, Chuo-Ku, Kobe, Hyogo 651-0086, Japan
e-mail: ozeki_akichika@lilly.com

K. Doksum
Department of Statistics, University of Wisconsin Madison, 1220 Medical Sciences Center, 1300 University Ave Madison, Madison, WI 53706, USA
e-mail: doksum@stat.wisc.edu

N. Hoshino et al. (eds.), *Pioneering Works on Extreme Value Theory*,
JSS Research Series in Statistics,
https://doi.org/10.1007/978-981-16-0768-4_4

where $X \in R$ is a covariate. We assume $\theta(x)$ is of the form $\theta(x) = g(x\beta)$, for some function g (the boundary structure). We develop efficient estimators of $\theta(x)$ for semiparametric models when the boundary structure is known. When the boundary structure is unknown, we propose a consistent estimator.

Such models with parameter-dependent support are irregular, because they do not satisfy the usual regularity conditions that much of statistical asymptotic theory is built on. A popular approach used for irregular models is Le Cam's asymptotic theory based on limits of experiments [18]. We apply the results reported by van der Vaart [25, 26], who gives a general development of the theory and applies it to the $UNIF[0, \theta]$ model. In this approach for parametric models, a local limit of a likelihood ratio provides a limit distribution that represents the limits of the experiments. By determining the optimal estimators of the parameters in this limit distribution, we obtain a lower bound on the asymptotic mean squared error (AMSE; or more generally, the asymptotic risk) of the estimators in the original experiment.

Estimators with an AMSE equal to the lower bound are said to be asymptotically efficient. We extend this approach to semiparametric models with an **unknown** parameter (θ, H), where $\theta \in R^d$ and H is a function. Here, we say estimators are semiparametrically adaptive and efficient if their asymptotic risk is equal to the lower bound on the asymptotic risk of the estimators in models in which the functional parameter H is **known**. This extension is used to develop semiparametrically adaptive and efficient estimators for boundary parameters.

Hirano and Porter [13] established a lower bound on the risk of parameter estimators in a parametric model, showing that a Bayes estimator is asymptotically efficient in the sense that the AMSE of the estimator reaches the lower bound. Other parametric or semiparametric methods for extremes include those of [7–9, 11, 12, 15, 22, 25, 27]. In economics, such boundary models are often referred to as parameter-dependent support models, and are used to study parametric auction models, search models, and production frontier models; see [13] and the references therein. Another approach is the quantile regression technique [17]. The quantile $F^{-1}(\tau = 1)$ (extremal quantile $\tau = 1$) is considered by [6, 16, 21, 24], although they do not examine efficiency. Here, $F(\cdot)$ is the distribution function of Y, $F(y) = P(Y \leq y)$, and $F^{-1}(\tau) = \inf\{y : F(y) \geq \tau\}$.

The extreme value theorem can also be used to estimate the upper boundary. As stated in [10], one of the most obvious estimators of the upper boundary for the one-sample case is the sample maximum. In fact, [14] (Remark 4.5.5) point out that using the sample maximum to estimate the boundary when the extreme value index satisfies $\xi < -1/2$ (our model has $\xi = -1$) is approximately equivalent to using the moment-related estimator for the endpoint. However, such a simple estimator produces a bias, and can be improved upon by using our approach. Another approach is to use the generalized Pareto distribution [23]. However, this approach converges more slowly than our approach does, and does not produce semiparametrically efficient estimators.

The remainder of the paper proceeds as follows. Section 4.2 introduces a boundary support model. Section 4.3 develops the concept of efficiency for both parametric and semiparametric models. The main result of the semiparametrically adaptive and efficient estimator when the boundary structure is known is described in Sect. 4.4 (see

[20] for the proof). Then, in Sect. 4.5, we propose a selection method for the boundary structure when this structure is unknown. Here, we also prove the consistency of the estimator. Real-data examples are presented in Sect. 4.6, and 4.7 concludes the paper.

4.2 Modeling Extremes

We start with a one-sample framework, where we assume that a random variable Y has an arbitrary continuous distribution function of the form $H(y)/H(\theta)$ on $[0, \theta]$, for $H(\cdot)$ continuous and increasing. Thus, we assume that $Y_1, \ldots, Y_n$ are independent and identically distributed (iid) as Y, with density

$$f(y;\theta) = \frac{h(y)}{H(\theta)} 1(0 \le y \le \theta), \qquad \theta \in \Theta = (0, \infty),\ H \in \mathcal{H}, \tag{4.1}$$

where $\mathcal{H} = \{H : H(y) = \int_0^y h(t)dt,\ h(\cdot)$ is a positive function on $[0, \infty)$, with $0 < h(0) < \infty\}$ and $\int_0^y h(t)dt < \infty$, for each $y < \infty$. The model given in (4.1) is a nonregular semiparametric model with parameter (θ, H). Here, H is unidentifiable because $H_1 \equiv H(y)$ and $H_2 \equiv cH(y)$, for $c > 0$, yield the same model (4.1); that is, $H_1 \ne H_2 \Rightarrow P_{H_1} = P_{H_2}$, for fixed θ. However, $f(y;\theta) = h(y)/H(\theta)$ is identifiable on $0 \le y \le \theta$.

When a covariate vector $X = (X_1, \ldots, X_d)^T$ is available, we model the conditional distribution of $Y|X = x$ as having the form (4.1), with

$$\theta = \theta(x) = g(x^T\beta),$$

where g is a known function, and $\beta \in R^d$ is a vector of *boundary regression parameters.* For example, a two-sample model can be modeled as $d = 2$ and $g(x^T\beta) = x_1\beta_1 + x_2\beta_2$, where (x_1, x_2) are dummy variables equal to (1, 0) and (0, 1) for sample 1 and sample 2, respectively.

4.3 Adaptive and Efficient Estimation

The usual results for efficient estimation do not apply to the nonregular model (4.1). It is nonregular in the sense that the density $f(y;\theta)$ is not differentiable in θ. Thus, we turn to asymptotic theory for such models. First, we introduce parametric models for which efficient estimators are defined. Next, we extend the parametric case to include semiparametric models and an "adaptive and efficient" estimator.

4.3.1 Parametric Models and Efficient Estimators

Let

$$P_\theta^n \equiv \text{ the distribution of } Y_1, \ldots, Y_n \text{ iid as } Y \sim P_\theta,$$

where P_θ is a distribution of $Y \in R$, and $\theta \in R^d$ is a parameter. Consider a collection of probability distributions

$$\mathcal{M} \equiv \{P_\theta^n : \theta \in \Theta \subseteq R^d\},$$

which serve as parametric models for the random vector $\boldsymbol{Y} = (Y_1, \ldots, Y_n)^T \in R^n$. When estimating a parameter $\psi \equiv \psi(\theta) : R^d \mapsto R$, an estimator $T \equiv T(\boldsymbol{Y}) : R^n \mapsto R$ is said to be c_n*-regular* (denoted as $T \in REG$) if there exists a sequence of positive numbers c_n, with $c_n \to \infty$ as $n \to \infty$, and a distribution Q_θ, such that for each $\theta \in \Theta$ and each sequence $\theta_n = \theta - \gamma/c_n$, for $\gamma \in R^d$,

$$c_n[T - \psi(\theta_n)] \overset{\theta_n}{\rightsquigarrow} Q_\theta, \tag{4.2}$$

where $\gamma/c_n \equiv (\gamma_1/c_n, \ldots, \gamma_d/c_n)$, Q_θ does not involve γ (denoted as $Q_\theta \perp \gamma$), and $\rightsquigarrow$ denotes weak convergence. Specifically, $\overset{\theta_n}{\rightsquigarrow}$ or $\overset{\gamma}{\rightsquigarrow}$ denotes weak convergence under $P_{\theta_n}^n$.

We define the risk of T as the AMSE

$$R(T; \psi) = E_{Q_\theta}[S^2],$$

where the random variable $S \sim Q_\theta$ represents the in-law limit of $c_n[T - \psi(\theta_n)]$ under $P_{\theta_n}^n$, defined in (4.2).

We say $T_1 \in REG$ is *asymptotically efficient* if

$$R(T_1; \psi) = \inf_{T \in REG} R(T; \psi).$$

Le Cam's approach to asymptotics is as follows: P_{θ_n} with $\theta_n = \theta - \gamma/c_n$ is a local approximation to P_θ, with parameter $\gamma = -c_n(\theta_n - \theta)$. Under certain conditions, there exists a probability distribution $P_\gamma^{(0)}$ such that the "distance" between $P_{\theta-\gamma/c_n}^n$ and $P_{\theta-\gamma_0/c_n}^n$ converges to the "distance" between $P_\gamma^{(0)}$ and $P_{\gamma_0}^{(0)}$. If we consider an experiment in which we draw one observation from $P_{\gamma_0}^{(0)}$, the optimal risk of the procedures for this experiment provides a lower bound for the asymptotic risk of the procedures in the original experiment with distribution P_θ^n. Our boundary parameter results carry over to general decision-theoretic procedures and risks.

Next, we examine this approach in greater detail. An experiment $\mathcal{E} = (\mathcal{Z}, \mathcal{A}, P_\gamma : \gamma \in \Gamma)$ is a collection of probability measures $\{P_\gamma : \gamma \in \Gamma\}$ on the sample space $(\mathcal{Z}, \mathcal{A})$. The experiment is interpreted as representing a model for a random variable Z on $(\mathcal{Z}, \mathcal{A})$, distributed as P_γ, for some γ that denotes a local parameter. This model is related to the original model by $\theta_n = \theta - \gamma/c_n$, for some fixed θ in the original parameter space.

A sequence of experiments $\mathcal{E}_n = (\mathcal{Z}_n, \mathcal{A}_n, P_{n,\gamma} : \gamma \in \Gamma)$ converges to a limit experiment $\mathcal{E} = (\mathcal{Z}, \mathcal{A}, P_\gamma^{(0)} : \gamma \in \Gamma)$ if the sequence of likelihood ratio processes converges marginally in distribution to the likelihood ratio process of the limit experiment. More precisely, define

$$P_{n,\gamma} \equiv P^n_{\theta-\gamma/c_n},$$

where $\theta \in R^d$ is a constant and $\gamma \in R^d$ is the parameter. The likelihood ratio for the local parameter γ converges to that of an experiment with one observation Z iff, for every finite subset $I \subset R^d$ and every $\gamma_0 \in R^d$, we have

$$\left(\frac{dP^n_{\theta-\gamma/c_n}}{dP^n_{\theta-\gamma_0/c_n}}(Y_1, \ldots Y_n)\right)_{\gamma\in I} \overset{\gamma_0}{\rightsquigarrow} \left(\frac{dP^{(0)}_\gamma}{dP^{(0)}_{\gamma_0}}(Z)\right)_{\gamma\in I},$$

where $Z \sim P^{(0)}_{\gamma_0}$. When the P has density p, we have

$$\frac{dP^n_{\theta-\gamma/c_n}}{dP^n_{\theta-\gamma_0/c_n}} = \frac{p^n_{\theta-\gamma/c_n}}{p^n_{\theta-\gamma_0/c_n}},$$

which makes the computation of the limit manageable.

Next, define a functional $\kappa(P_{n,\gamma}) \equiv \psi(\theta_n)$, where $\psi(\theta_n)$ is given in (4.2). We say $\kappa(P_{n,\gamma})$ is *differentiable in the limit with derivative* $\kappa'(\gamma)$ if there exists $\kappa'(\gamma)$ such that, as $n \to \infty$,

$$c_n[\kappa(P_{n,\gamma}) - \kappa(P_{n,\gamma_0})] \overset{n}{\to} \kappa'(\gamma) - \kappa'(\gamma_0). \tag{4.3}$$

Using [25] (Theorems 3.1, 4.1, and 5.1), we have the following theorem.

Theorem 4.1 *Suppose $T \in REG$ and (4.3) holds, and let Z be a random variable with distribution $P^{(0)}_\gamma$, where we have simplified the notation by writing γ instead of γ_0. Assume that the distribution of Z can be written as the distribution of $V + \Sigma\gamma$, for some nonsingular matrix $\Sigma \in R^{d\times d}$ and some random variable V on R^d that has an absolutely continuous distribution that does not involve γ. Write $\mathcal{L}$ for "law" or "distribution of."*

Then, the limit distribution of T in (4.2) can be written as

$$Q_\theta = \mathcal{L}(S) = \mathcal{L}(\kappa'(\Sigma^{-1}V) + W), \tag{4.4}$$

where W is a random variable on R independent of V.

In (4.4), we can regard W as a noise variable that contributes to the risk. We get rid of W by noting that for each W in (4.4), $E_{Q_\theta}[S^2]$ is bounded below by $Var_{Q_\theta}[S^2]$.

Corollary 4.1

$$\inf_{T \in REG} R(T; \psi) \geq E[\kappa'(\Sigma^{-1}V) - E\kappa'(\Sigma^{-1}V)]^2.$$

4.3.2 Semiparametric Models and Adaptive and Efficient Estimators

Here, we extend the theory in Sect. 4.3.1 to include semiparametric models of the form

$$\mathcal{M} = \{P^n(\cdot|\theta, H) : \theta \in \Theta \subseteq R^d, H \in \mathcal{H}\},$$

where $\boldsymbol{Y} = (Y_1, \ldots, Y_n)^T \sim P^n(\cdot|\theta, H), Y_1, \ldots, Y_n$ iid as $Y \sim P(\cdot|\theta, H)$, and $\mathcal{H}$ is a class of functions satisfying regularity conditions that are specified for each specific model considered. The goal is to estimate

$$\psi \equiv \psi(\theta) : R^d \mapsto R.$$

Assume, temporarily, that H is known, and let $R(T; \psi|H)$ denote the AMSE of the estimator T of ψ, where $R(T; \psi|H)$ can involve H. From the parametric model discussion, we have

$$R(T; \psi|H) = E_{Q_\theta}[S^2],$$

where the random variable S represents the in-law limit of $c_n[T - \psi(\theta_n)]$ under $P^n(\cdot|\theta_n, H)$, with $\theta_n = \theta - \gamma/c_n$. Here, $T \in REG$ and Q_θ are defined as in (4.2), where Q_θ does not involve γ, but may involve H.

We say that $\hat{\psi}$, an estimator of $\psi(\theta)$, is *semiparametrically adaptive and efficient* (see [1, 3] (Sect. 6.2.2), and [4] for adaptive estimation) if $\hat{\psi} \in REG$, $\hat{\psi}$ does not involve H, and $\hat{\psi}$ reaches the asymptotic lower bound on the risk for the case of known H; that is,

$$R(\hat{\psi}; \psi) = \inf_{T \in REG} R(T; \psi|H).$$

Note that $R(\cdot; \psi)$ is the AMSE risk when H is unknown, and $R(\cdot; \psi|H)$ is the AMSE risk when H is known. In other words, the estimation of ψ under a semiparametric model (H unknown) can be performed as though we know H.

4.4 Semiparametric Boundary Parameter Regression with a Known Boundary Structure

In this section, we introduce a semiparametrically adaptive and efficient estimator. Consider a random design for the response variable Y_i, where (X_i, Y_i) are iid as (X, Y), with conditional density

$$f(y|x;\beta) = \frac{h(y)}{H(g(x\beta))} \cdot 1(0 \leq y \leq g(x\beta)), \tag{4.5}$$

and a cumulative distribution function (CDF)

$$F(y|x;\beta) = \begin{cases} \frac{H(y)}{H(g(x\beta))} & (0 \leq y \leq g(x\beta)) \\ 1 & (g(x\beta) < y). \end{cases} \tag{4.6}$$

Here, the boundary parameter $\beta \in R$ is unknown, H is unknown, and the boundary structure $g(\cdot)$ is known.

We assume the following regularity conditions:

$[C1]$ $x \in \Omega_X \equiv [a, b]$, for $0 < a < b < \infty$, the density of X, $f_X(x) > 0$ on Ω_X.
$[C2]$ $\beta \in \Omega_\beta \equiv (a_1, b_1)$, for $0 < a_1 < b_1 < \infty$, the second derivative $d^2h(y)/dy^2 \equiv h''$ is continuous and bounded on $[0, \infty)$, and $0 < \varepsilon < \inf_{y>0} h(y)$, for some $\varepsilon > 0$.
$[G1]$ $0 < \inf_{\beta\in\Omega_\beta, x\in\Omega_X} g(x\beta)$.
$[G2]$ $0 < \inf_{y>0} g'(y)$; that is, $g(\cdot)$ is strictly increasing, where $dg(y)/dy = g'(y)$.
$[G3]$ $d^3g(x\beta)/d\beta^3 \equiv g'''$ is continuous in β and $\sup|g'| < \infty$, $\sup|g''| < \infty$, and $\sup|g'''| < \infty$ on the set $\{\beta \in \Omega_\beta, x \in \Omega_X\}$, where g' and g'' are first and second derivatives, respectively, of g with respect to β.
$[G4]$ The inverse function satisfies $g^{(-1)}(0) = 0$.

Theorem 4.2 *Assume the semiparametric model (4.6) (known boundary structure g) and the regularity condition above. Using Le Cam's limits of experiments approach, we can construct an n-regular $\hat{\beta}$ with an AMSE that achieves the lower bound*

$$R(\hat{\beta};\beta) = \inf_{T\in REG} R(T;\beta|H) = \left(E_X\left[\frac{h(g(\beta X))g'(\beta X)X}{H(g(\beta X))}\right]\right)^{-2}.$$

Hence, $\hat{\beta}$ is semiparametrically adaptive and efficient.

Proof See [20]. □

The estimator $\hat{\beta}$ is constructed as follows: Set

$$V_i = \frac{g^{(-1)}(Y_i)}{X_i}, \quad i = 1, \ldots, n. \tag{4.7}$$

Then, $V_1, \ldots, V_n$ are iid. Let $f_V(v;\beta)$ and $F_V(v;\beta)$ denote the density of V_i and the CDF, respectively. Note that f_V has support $[0, \beta]$. This is a boundary support model. Therefore, we can use the framework described in Sects. 4.2 and 4.3 with Y_i replaced with V_i and θ replaced with β. Let $V_{(n)}$ be the largest order statistic. Then, $V_{(n)}$ is a negatively biased estimator of β. Ozeki [20] found that

$$V_{(n)} + \frac{1}{nf_V(\beta;\beta)} \tag{4.8}$$

is asymptotically unbiased. Therefore, a consistent estimator of the density of V at the right boundary β is an asymptotically unbiased estimator of β.

Viewing $f_V(\beta;\beta)$ as the derivative of the distribution function $F_V(v;\beta)$ at the boundary leads to the estimator

$$\begin{aligned}\hat{f}_\delta &\equiv \frac{n^2}{n^2-1}\cdot\frac{\hat{F}(V_{(n)})-\hat{F}(V_{(n)}-\delta)}{\delta}\\ &= \frac{n}{n^2-1}\frac{1}{\delta}\left[\sum_{i=1}^{n-1} 1(V_{(n)}-\delta < V_{(i)})+1\right],\end{aligned} \tag{4.9}$$

where $\hat{F}$ is the empirical distribution of $V_1,\ldots,V_n$, and δ is a small positive constant that depends on n. By investigating the asymptotic distribution of $\hat{f}_\delta$, we find that $\hat{f}_\delta$ is a consistent estimator of $f_V(\beta;\beta)$, provided that $\delta\to 0$ and $n\delta\to\infty$ as $n\to\infty$. In addition, we find that the optimal choice of δ is $\delta_0 = cn^{-1/3}$, where the constant c depends on f_V. In the simulation studies described in [20], $c=1$ is identified as a good choice. We use this setting in Sect. 4.6.

From (4.8) and (4.9), we have the following estimator:

$$\hat{\beta} \equiv V_{(n)} + \frac{1}{n\hat{f}_\delta}. \tag{4.10}$$

This estimator is n-regular, $\hat{\beta} \rightsquigarrow \beta$, and it achieves the AMSE lower bound, as stated in Theorem 4.2, provided that $\delta\to 0$ and $n\delta\to\infty$ as $n\to\infty$. See [20] for a detailed discussion of the properties of the maximum likelihood estimator (MLE) $\hat{\beta}_{MLE} \equiv V_{(n)}$.

4.5 Semiparametric Boundary Parameter Regression with an Unknown Boundary Structure

In this section, we propose a consistent estimator for a semiparametric boundary parameter regression with an unknown boundary structure under regularity conditions. Note that efficiency is not proved.

Sometimes, it is too optimistic to assume that the boundary structure $g(\cdot)$ is known. Consider a more general boundary structure $g(x,\beta)$. Then, the conditional density (4.5) and the CDF (4.6) become

$$f(y|x;\beta) = \frac{h(y)}{H(g(x,\beta))}1(0\le y\le g(x,\beta)), \tag{4.11}$$

and

$$F(y|x;\beta) = \begin{cases} \frac{H(y)}{H(g(x,\beta))} & (0 \le y \le g(x,\beta)) \\ 1 & (g(x,\beta) < y), \end{cases} \tag{4.12}$$

respectively. Suppose the true boundary structure g is unknown, but that it lies in a certain parametric class $g \in \mathcal{G}$. Then, the following criteria can be used to identify the best model within the class:

$$FitTest = \min_{m \in \mathcal{G}} \sum_{i}^{n} m(X_i, \hat{\beta}_m), \qquad \text{such that } m(X_i, \hat{\beta}_m) \ge Y_i \text{ for all } i,$$

$$\hat{g} = \arg\min_{m \in \mathcal{G}} \sum_{i=1}^{n} m(X_i, \hat{\beta}_m), \qquad \text{such that } m(X_i, \hat{\beta}_m) \ge Y_i \text{ for all } i, \tag{4.13}$$

where $\hat{\beta}_m$ is in (4.10) for the model with the boundary structure $m(\cdot,\cdot)$. More specifically, $m(\cdot,\cdot)$ is the boundary structure if the model is based on (4.11), where $g(\cdot,\cdot)$ is replaced with $m(\cdot,\cdot)$; $\hat{g}$ minimizes the FitTest in (4.13). Our final estimator is

$$\hat{\beta}_{\hat{g}}. \tag{4.14}$$

The FitTest tries to find the smallest boundary m that encloses all Y. This is related to the methods proposed by [2, 5], who introduce a model parameter λ and use a maximum likelihood or robust estimation approach to estimate β and λ. Thus, we assume that the boundary structure is of the form $g_\lambda(x,\beta)$, and that the data are generated using model (4.11), where g is replaced with g_λ. Here, (β_0, λ_0) denotes the true model parameters. In this setting, the likelihood is proportional to

$$L(\beta,\lambda) = \prod_{i=1}^{n} \frac{h(Y_i) \cdot 1(0 \le Y_i \le g_\lambda(X_i,\beta))}{H(g_\lambda(X_i,\beta))}.$$

In the approach of [5], they first fix λ and find the "MLE" $\hat{\beta}(\lambda)$ of β as

$$\hat{\beta}(\lambda) = \arg\max_{\beta} L(\beta,\lambda).$$

We refer to this estimator as a profile-MLE (PMLE), and denote it as $\hat{\beta}_{m,PMLE}$ in (4.19). This leads to the profile likelihood

$$L(\lambda) = \prod_{i=1}^{n} \frac{h(Y_i) \cdot 1(0 \le Y_i \le g_\lambda(X_i,\hat{\beta}(\lambda)))}{H(g_\lambda(X_i,\hat{\beta}(\lambda)))}.$$

In our case, let $\hat{\beta}(\lambda)$ be defined as in (4.10), with g_λ used in place of g. The next step is to let

$$\hat{\lambda} = \arg\max_{\lambda} L(\lambda) = \arg\max_{\lambda} \log L(\lambda) = \arg\max_{\lambda} \sum_i \log \frac{1(0 \le Y_i \le g_\lambda(X_i, \hat{\beta}(\lambda)))}{H(g_\lambda(X_i, \hat{\beta}(\lambda)))}$$

$$= \arg\min_{\lambda} \sum_i \log H(g_\lambda(X_i, \hat{\beta}(\lambda))), \quad \text{such that } g_\lambda(X_i, \hat{\beta}(\lambda)) \ge Y_i \text{ for all } i, \tag{4.15}$$

yielding the final estimator

$$\hat{\beta} = \hat{\beta}(\hat{\lambda}).$$

The properties of these profile estimators have been studied extensively for regular models; see [2, 5], among others. However, the properties of nonregular models remain an interesting open problem. In practice, we can restrict λ to be in a discrete set of values $\{\lambda_1, \dots, \lambda_K\}$ to ease the computation burden.

Note that the profile likelihood approach in (4.15) has an additional monotone transformation $\log H(\cdot)$ in the objective function to those of the FitTest in (4.13). The profile likelihood approach is difficult because it involves the unknown $H(\cdot)$. We prove that the best model selected by the FitTest is the true model as $n \to \infty$ when $\mathcal{G}$ is "small"; this is stated as the regularity condition $[M7]$ below.

Let $m(\cdot, \cdot), g(\cdot, \cdot) \in \mathcal{G}$, where g is the true model and m is a wrong model (i.e., a misspecification). Define the inverse function of $y = m_x(\beta)$ as $m_x^{(-1)}(y) = \beta$. For simplicity, we may write $m(x, \beta) = m_x(\beta)$, $\partial m_x(v)/\partial v = m_x'(v)$, $\partial^2 m_x(v)/\partial v^2 = m_x''(v)$, and $\partial^3 m_x(v)/\partial v^3 = m_x'''(v)$.

Define

$$\sup_{x \in \Omega_X} m_x^{(-1)}(g_x(\beta)) \equiv B_m, \tag{4.16}$$

and the following regularity conditions of $\mathcal{G}$:

$[M1]$ $\inf_{x \in \Omega_X, \beta \in \Omega_\beta} m_x(\beta) > 0$.
$[M2]$ $\inf_v m_x'(v) > 0$. Here, inf is taken over $\{v \in [0, B_m]; \beta \in \Omega_\beta, x \in \Omega_X\}$, where B_m is defined in (4.16); $m(x, \beta)$ is a strictly increasing continuous function in x and in β.
$[M3]$ $m_x'''(v)$ is continuous in v and $\sup |m_x'(v)| < \infty$, $\sup |m_x''(v)| < \infty$, and $\sup |m_x'''(v)| < \infty$. Here, sup is taken over $\{v \in [0, B_m], x \in \Omega_X; \beta \in \Omega_\beta\}$, where B_m is defined in (4.16).
$[M4]$ $m_x^{(-1)}(0) = 0$.
$[M5]$ $m_x^{(-1)}(y)$ is continuous in x.
$[M6]$ $P(m_X(\beta_1) \ne g_X(\beta)) = 1$, for any $\beta_1 \in \Omega_\beta$.
$[M7]$ $\mathcal{G}$ is a finite set. That is, $\mathcal{G} = \{g, m_1, m_2, \dots, m_K\}$, for some positive integer K.

Note that $[M1]$, $[M2]$, $[M3]$, and $[M4]$ contain the true model g's regularity conditions $[G1]$, $[G2]$, $[G3]$, and $[G4]$, respectively, in Sect. 4.4.

Define a transformed random variable under the model misspecification as

$$Z_i = m_{X_i}^{(-1)}(Y_i), \quad i = 1, \dots, n.$$

This is the counterpart of the transformation under the true boundary structure $g(\cdot)$ in (4.7). Let $Z_1, \ldots, Z_n$ be iid as Z. After some algebra with (4.11) and (4.12), the density and the CDF of Z are shown to be

$$f_Z(v;\beta) = E_X\left[\frac{h(m_X(v))m'_X(v)}{H(g_X(\beta))}\cdot 1(0\le v\le m_X^{(-1)}(g_X(\beta)))\right],$$
$$F_Z(v;\beta) = E_X\left[\left.\frac{H(m_X(v_0))}{H(g_X(\beta))}\right|_{v_0=\min\{v,m_X^{(-1)}(g_X(\beta))\}}\right],$$

respectively. The supremum support of the random variable Z is defined as B_m in (4.16). We have

$$|B_m| < \infty, \tag{4.17}$$

from $[C1]$, $[M2]$, and $[M5]$, which guarantees $m_x^{(-1)}(g_x(\beta))$ is continuous on the compact set $x \in \Omega_X$. Note that the support of $V = g_X^{(-1)}(Y)$ is $[0, \beta]$ with $F_V(\beta;\beta) = 1$, and the support of $Z = m_X^{(-1)}(Y)$ is $[0, B_m]$ with $F_Z(B_m;\beta) = 1$. We assume $B_m \in \Omega_\beta$, without loss of generality.

Now, define the best estimator under the wrong model m. Based on (4.9), we have the boundary density estimator

$$\begin{aligned}\hat{f}_{m,\delta} &\equiv \frac{n^2}{n^2-1}\cdot\frac{\hat{F}_Z(Z_{(n)}) - \hat{F}_Z(Z_{(n)}-\delta)}{\delta}\\ &= \frac{n}{n^2-1}\frac{1}{\delta}\left[\sum_{i=1}^{n-1} 1(Z_{(n)}-\delta < Z_{(i)}) + 1\right],\end{aligned} \tag{4.18}$$

for $f_Z(B_m;\beta)$. Here, $\hat{F}_Z$ is the empirical distribution of Z_i, for $i = 1, \ldots, n$, and $\delta = cn^{-1/3}$.

Lemma 4.1

$$\frac{1}{n\hat{f}_{m,\delta}} \rightsquigarrow 0.$$

Proof From (4.18), we have

$$\hat{f}_{m,\delta} \ge \frac{n}{n^2-1}\frac{1}{\delta}(0+1) = \frac{n}{n^2-1}\cdot\frac{1}{cn^{-1/3}},$$

with probability one. Hence,

$$\frac{1}{n\hat{f}_{m,\delta}} \le \frac{n^2-1}{n^2}\frac{c}{n^{1/3}} \rightsquigarrow 0. \qquad \square$$

Next, observe that $F_Z(v;\beta) < 1$ for $v < B_m$, because $h(\cdot)$ and $m'_x(\cdot)$ in the expectation are strictly positive on the support of v, from [$C1$] and [$M2$]. Hence using a well-know result, we can show the convergence

$$\hat{\beta}_{m,PMLE} \equiv Z_{(n)} \to B_m, \tag{4.19}$$

almost surely.

Considering Lemma 4.1 with Slutsky's Theorem, Lemma 4.2 shows the asymptotics of the efficient estimator under the wrong model m.

Lemma 4.2

$$\hat{\beta}_m \equiv \hat{\beta}_{m,PMLE} + \frac{1}{n\hat{f}_{m,\delta}} \rightsquigarrow B_m. \tag{4.20}$$

Note that $\hat{\beta}$ in (4.10) (under the true model) is a special case of $\hat{\beta}_m$ in (4.20).

Lemma 4.3

$$P(m_X(B_m) > g_X(\beta)) = 1.$$

Proof Use contradiction. Because $m_x(\cdot)$ is a wrong model, there exists an $x_0 \in \Omega_X$ such that $m_{x_0}(B_m) \neq g_{x_0}(\beta)$, by [$M6$]. Suppose $m_{x_0}(B_m) < g_{x_0}(\beta)$. Because $m_{x_0}(B_m)$ and $g_{x_0}(\beta)$ are continuous in x_0, by [$M2$], there exist constants a', b', where $0 < a \le a' < x_0 < b' \le b < \infty$ and

$$0 < \inf_{x\in[a',b']} (g_x(\beta) - m_x(B_m)).$$

Define the area

$$A_{(x,y)} \equiv \{(x, y) : m_x(B_m) \le y \le g_x(\beta), a' \le x \le b'\}.$$

Then, by (4.11), [$C1$], and [$C2$],

$$P_{X,Y}((X, Y) \in A_{(x,y)}) = \int_{a'}^{b'} \int_{m_x(B_m)}^{g_x(\beta)} f(y|x,\beta)dy dP_X(x) > 0.$$

On the other hand, by definition, $P(Y \le m_X(B_m)) = 1$, which is a contradiction. □

Lemma 4.4 *As* $n \to \infty$,

$$P\left(\frac{1}{n}\sum_{i=1}^{n}(m_{X_i}(\hat{\beta}_m) - g_{X_i}(\hat{\beta})) > 0\right) \to 1. \tag{4.21}$$

Proof For simplicity, write $m_{X_i}(\hat{\beta}_m) \equiv \hat{m}_i$, $g_{X_i}(\hat{\beta}) \equiv \hat{g}_i$, $m_{X_i}(B_m) \equiv m_i$, $g_{X_i}(\beta) \equiv g_i$, $\partial m_{X_i}(v)/\partial v \equiv m'_i(v)$, and $\partial g_{X_i}(v)/\partial v \equiv g'_i(v)$. Then, (4.21) becomes

$$\frac{1}{n}\sum_{i=1}^{n}(\hat{m}_i - \hat{g}_i) = \frac{1}{n}\sum_{i=1}^{n}(\hat{m}_i - m_i) + \frac{1}{n}\sum_{i=1}^{n}(m_i - g_i) + \frac{1}{n}\sum_{i=1}^{n}(g_i - \hat{g}_i) \equiv K1 + K2 + K3. \tag{4.22}$$

For $K1$ and $K3$, Taylor's Theorem yields

$$\hat{m}_i = m_i + m_i'(B_i^*)(\hat{\beta}_m - B_m),$$
$$\hat{g}_i = g_i + g_i'(\beta_i^*)(\hat{\beta} - \beta),$$

where B_i^* is between $\hat{\beta}_m$ and B_m, and β_i^* is between $\hat{\beta}$ and β. We have

$$\begin{aligned}
|K1| &\le \left|\hat{\beta}_m - B_m\right| \left|\frac{\sum m_i'(B_i^*)}{n}\right| \le \left|\hat{\beta}_m - B_m\right| \left|\frac{\sum \sup_{x\in\Omega_X,\beta\in\Omega_\beta} m_x'(\beta)}{n}\right| \\
&\le |\hat{\beta}_m - B_m| \left|\frac{\sum G_1}{n}\right| = |\hat{\beta}_m - B_m| \cdot G_1, \\
|K3| &\le \left|\hat{\beta} - \beta\right| \left|\frac{\sum g_i'(\beta_i^*)}{n}\right| \le \left|\hat{\beta} - \beta\right| \left|\frac{\sum \sup_{x\in\Omega_X,\beta\in\Omega_\beta} g_x'(\beta)}{n}\right| \\
&\le |\hat{\beta} - \beta| \left|\frac{\sum G_2}{n}\right| = |\hat{\beta} - \beta| \cdot G_2,
\end{aligned}$$

from $[M3]$ and (4.17). Here, G_1, G_2 are positive constants. By Lemma 4.2, $|K1| = o_P(1)$ and $|K3| = o_P(1)$. Here, $o_P(1)$ means convergence in probability to zero.

For $K2$, the strong law of large numbers and Lemma 4.3 give almost sure convergence,

$$K2 = \frac{1}{n}\sum_{i=1}^{n}(m_i - g_i) \to E_X[m_X(B_m) - g_X(\beta)] > 0.$$

Here, the expectation is finite, by $[C1]$ and $[M2]$. Hence, by Slutsky's Theorem, (4.22) becomes

$$\frac{1}{n}\sum_{i=1}^{n}(\hat{m}_i - \hat{g}_i) = K1 + K2 + K3 \rightsquigarrow E_X[m_X(B_m) - g_X(\beta)] > 0.$$

This proves (4.21).

Theorem 4.3 combines these results.

Theorem 4.3 *Consider a semiparametric boundary support model with distribution function (4.12), where the boundary structure $g(\cdot,\cdot)$ is unknown, but belongs to a parametric class $\mathcal{G}$. Suppose the regularity conditions $[C1]$, $[C2]$, and $[M1]$-$[M7]$ hold. Define the estimator $\hat{\beta}_m$ under a model $m \in \mathcal{G}$ as in (4.20). As $n \to \infty$, the*

FitTest (4.13) chooses the true boundary model g with probability tending to one. Moreover, the estimator (4.14) is consistent,

$$\hat{\beta}_{\hat{g}} \rightsquigarrow \beta.$$

Proof Consider the minimization problem (4.13). From Lemma 4.4, for $m_k \in \mathcal{G}$, $k = 1, \ldots, K$, by $[M7]$, we have for any $0 < \varepsilon_1$, there exists N_0 such that for all $N_0 < n$ and all $k \in \{1, \ldots, K\}$,

$$P\left(\sum_{i=1}^{n} m_k(X_i, \hat{\beta}_{m_k}) \leq \sum_{i=1}^{n} g(X_i, \hat{\beta})\right) \equiv P(S_k) \leq \varepsilon_1.$$

By De Morgan's Laws, we have

$$\begin{aligned} \{\hat{g} = g\} &= \left\{\sum_{i=1}^{n} m_k(X_i, \hat{\beta}_{m_k}) > \sum_{i=1}^{n} g(X_i, \hat{\beta}), \quad \text{for all } k \in \{1, \ldots, K\}\right\} \\ &= \overline{S_1} \cap \cdots \cap \overline{S_K}, \\ \{\hat{g} \neq g\} &= \cup_{k=1}^{K} S_k. \end{aligned}$$

Then, for all $N_0 < n$, by Boole's Inequality,

$$P(\hat{g} \neq g) = P(\cup_{k=1}^{K} S_k) \leq \sum_{k=1}^{K} P(S_k) = K\varepsilon_1.$$

By setting $\varepsilon_2/2 = K\varepsilon_1$, we have

$$P(\hat{g} \neq g) \leq \varepsilon_2/2.$$

In addition, by Lemma 4.2 (under the true model), for any $0 < \varepsilon$, there exists N_g such that for all $N_g < n$, $P(|\hat{\beta} - \beta| > \varepsilon) \leq \varepsilon_2/2$. Let $N \equiv \max(N_0, N_g)$. Then, for all $N < n$, we have

$$\begin{aligned} P(|\hat{\beta}_{\hat{g}} - \beta| > \varepsilon) &= P(|(\hat{\beta}_{\hat{g}} - \hat{\beta}) + (\hat{\beta} - \beta)| > \varepsilon, \{\hat{g} = g\}) \\ &\quad + P(|(\hat{\beta}_{\hat{g}} - \hat{\beta}) + (\hat{\beta} - \beta)| > \varepsilon, \{\hat{g} \neq g\}) \\ &\leq P(|0 + (\hat{\beta} - \beta)| > \varepsilon, \{\hat{g} = g\}) + \varepsilon_2/2 \\ &\leq \varepsilon_2/2 + \varepsilon_2/2 = \varepsilon_2. \end{aligned}$$

Hence $\hat{\beta}_{\hat{g}}$ is consistent. □

4.6 Real-Data Analysis

4.6.1 Example 1: Survival Time of Lung Cancer Patients (Two-Sample Models)

The data describe the survival of patients with advanced lung cancer, and are provided by the North Central Cancer Treatment Group (R package survival.cancer [19]). Here, we examine whether a significant difference exists between males and females in terms of longevity (the longest survival time). We fit a semiparametric two-sample boundary support model as a special case of the semiparametric regression model in which two independent samples of nonnegative responses, $X_1, \ldots, X_m$ iid as X (male), and $Z_1, \ldots, Z_n$ iid as Z (female), have densities

$$f_1(x;\theta_1) = \frac{h_1(x)}{H_1(\theta_1)} \cdot 1(0 \le x \le \theta_1),\ f_2(z;\theta_2) = \frac{h_2(z)}{H_2(\theta_2)} \cdot 1(0 \le z \le \theta_2),$$

where $H_j(\cdot)$, for $j = 1, 2$, are unknown continuous increasing functions on $R^+ = [0, \infty)$, with derivatives $h_j(\cdot)$. To compare the extremes of the X and Z populations, we consider the *shift parameter* ς,

$$\varsigma = \theta_1 - \theta_2.$$

We estimate the difference (male–female) of the extreme quantiles. Figure 4.1 shows the density plots for 112 males and 53 females. Except for the right tail, the males density is shifted left compared with that of the females. The summary statistics are as follows: first quartile (118, 167), mean survival time (263, 326), median (209, 293), third quartile (363, 444), and maximum (883, 765), where the values in parentheses represent males and females, respectively.

Based on the adaptive and efficient estimator in Theorem 4.2, the maximum survival time is estimated as 910 days for males and 782 days for females. Furthermore, we apply the adaptive and efficient estimator for the difference between the two. The male extreme survival is 127 days longer than that of females, with 95% confidence interval = (75, 205). Thus, there is a significant difference between males and females in terms of longevity. Note that the confidence interval is based on Theorem 4.1; see [20] for further details.

4.6.2 Example 2: Survival Time of Lung Cancer Patients (calorie)

We use the same data as in Example 1. Loprinzi et al. [19] shows that caloric intake is correlated with patient survival; that is, appetite is an indicator of how well a patient is

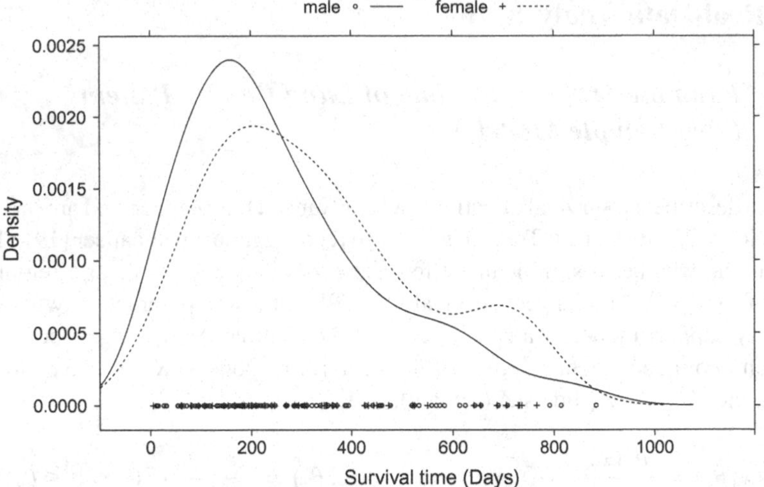

Fig. 4.1 Density plots for male and female survival time

doing. We are interested in how a patient's caloric intake relates to his/her survival at the extreme quantile $\tau = 1$; that is, $F^{-1}(\tau = 1) = g(x, \beta)$, where $F^{-1}(\tau) = \inf\{y : F(y|x; \beta) \geq \tau\}$. Here, F is a distribution function in (4.12). In other words, we want to estimate the longest survival time Y (in days), given a certain meal intake X (in calories). We use patients who have both survival time and calorie data ($n = 134$). As a preliminary, we fit a simple linear regression (Fig. 4.2). The straight line is the regression line, and the shaded area is the 95% predictive interval. Figure 4.2 suggests a positive trend. That is, on average, increased calorie intake is associated with a longer survival time.

We fit a semiparametric boundary support model with one covariate, and apply Theorem 4.3. In this model, we assume the boundary structure is unknown, but that it belongs to the class

$$\mathcal{G} = \{g(x, \beta) = x^C \beta; \quad C \in \{1/10, 1/5, 1/2, 1, 2\}\}.$$

We assume the true boundary structure $g(\cdot, \cdot)$ lies within this class. We use the criteria (4.13) to identify the best model. The regularity conditions of $\mathcal{G}$ in Theorem 4.3 can be checked as follows.

Define the true boundary structure $g(x, \beta)$ and a wrong model $m(x, \beta)$, and its related functions, as follows:

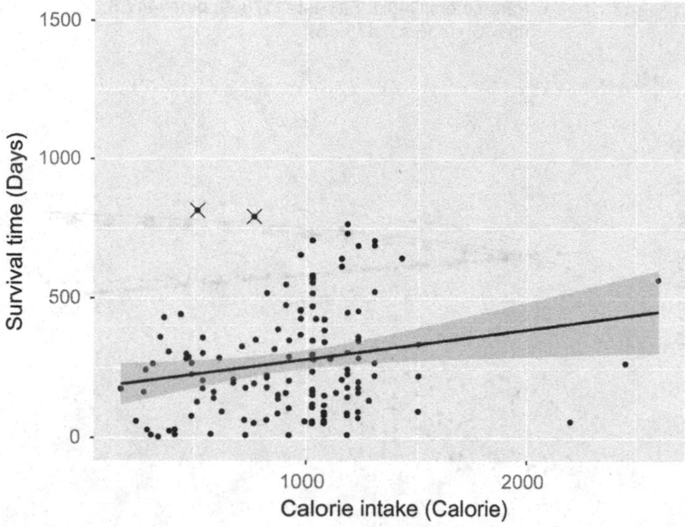

Fig. 4.2 Survival time of lung cancer patients. The straight line is a linear regression with a shaded 95% prediction interval. Patients with the longest and second-longest survival times (calories, time) = (513, 814), (768, 791) are denoted by an $\times$

$$
\begin{aligned}
g(x, \beta) &= x^{C_0}\beta, \\
m(x, \beta) &= m_x(\beta) = x^C\beta, \\
m_x^{(-1)}(y) &= y/x^C, \\
m_x^{(-1)}(g_x(\beta)) &= x^{C_0 - C}\beta, \\
m_x'(v) &= x^C, \\
m_x''(v) &= m_x'''(v) = 0.
\end{aligned}
$$

We assume $[C1]$ $x \in \Omega_X \equiv [1, 3000]$, and $[C2]$ $\beta \in \Omega_\beta \equiv (a_1, b_1)$, for $0 < a_1 < b_1 < \infty$. We assume the other conditions in $[C2]$. $[M1]$–$[M5]$ can be easily checked. For example, in $[M5]$, $m_x^{(-1)}(y) = y/x^C$ is continuous in $x \in \Omega_X$. For $[M6]$, we solve the equation $x^C\beta_1 = x^{C_0}\beta$ with respect to β_1. This yields $\beta_1 = x^{-C+C_0}\beta$. Therefore, when $C = C_0$, we have an identifiability problem. Any combination of sets in $\mathcal{G}$ can clear this condition. Lastly, $[M7]$ is $K = 4$ (including the true g, we have five elements in $\mathcal{G}$.)

We estimate the extreme quantile using four methods: the consistent estimator $\hat{\beta}_{\hat{g}}$ in Theorem 4.3 (95% confidence interval); $\hat{\beta}_{\hat{g},PMLE}$ (PMLE under model $\hat{g}$ in (4.19); 0.99 quantile regression (original scale); and 0.99 quantile regression on the transformed scale (quantile regression on given C) (see Fig. 4.3).

Here, the 95% confidence interval of $\hat{\beta}_{\hat{g}}$ is based on the asymptotics. We can show by Theorems 4.2 and 4.3 that on the set $\{\hat{g} = g\}$,

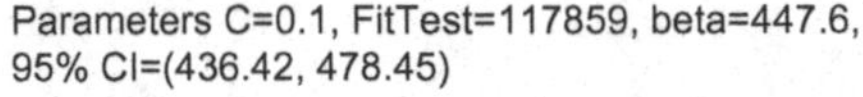

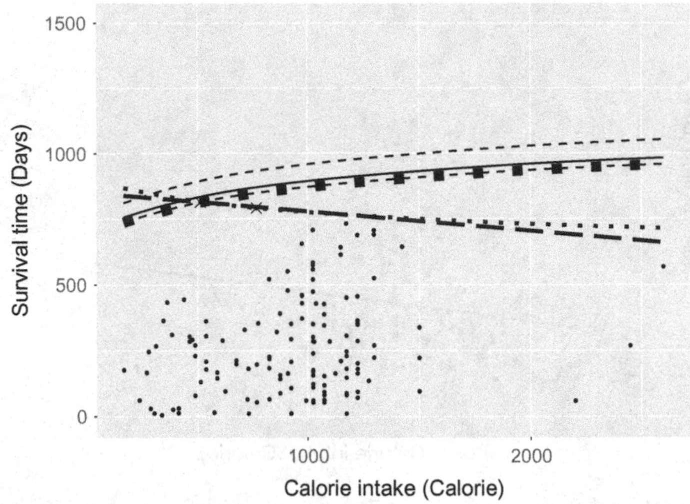

Fig. 4.3 Two estimators for the $\tau = 0.99$ quantile regression (long-dash line from the usual quantile regression, dotted-line curve from the quantile regression on the transformed scale [17]), and two estimators for $\tau = 1$ ($\hat{g}(X, \hat{\beta}_{\hat{g},PMLE}) = X^{0.1}\hat{\beta}_{\hat{g},PMLE}$, with a dotted diamond, and $\hat{g}(X, \hat{\beta}_{\hat{g}}) = X^{0.1}\hat{\beta}_{\hat{g}}$ with a solid-line curve (dashed 95% C.I.)); $\hat{\beta}_{\hat{g}}$ is based on the FitTest in (4.14)

$$-n(\hat{\beta}_{\hat{g}} - \beta) = -n(\hat{\beta}_g - \beta) \rightsquigarrow W - \alpha, \tag{4.23}$$

where $W \sim EXP[0, \alpha]$ (exponential distribution with shift $= 0$, scale $= \alpha$), with $\alpha = 1/f_V(\beta; \beta)$. Using (4.18), we substitute $\hat{\alpha} = 1/\hat{f}_{\hat{g},\delta}$ into the right-hand side (RHS) of (4.23). Then, we generate random variables W_1, 10^7 times, where $W_1 \sim EXP[0, \hat{\alpha}]$. Then, we compute the lower and upper 2.5% quantiles,

$$\beta_L = \hat{\beta}_{\hat{g}} + \text{(RHS of (4.23) lower 2.5\% quantile)/n},$$
$$\beta_U = \hat{\beta}_{\hat{g}} + \text{(RHS of (4.23) upper 2.5\% quantile)/n}.$$

For the 0.99 quantile regression on the transformed scale, we have the boundary condition $\{y \le g(x, \beta)\} = \{y \le x^C\beta\}$. This leads to a transformed model, defined as the response variable Y and a predictor variable X^C. We fit the quantile regression using an intercept and X^C as covariates. The quantile regression on the original scale (the response Y and the predictor variable X) is a straight line. However, the quantile regression on the transformed scale is a curve, because it is transformed back to the original scale. On the semiparametric boundary support model, the best model is selected as $C = 0.1$. That is, the best model is $\hat{g}(X, \beta) = X^{0.1}\beta$, with $FitTest = 117859$ and $\hat{\beta}_{\hat{g}} = 447.6(436.42, 478.45)$.

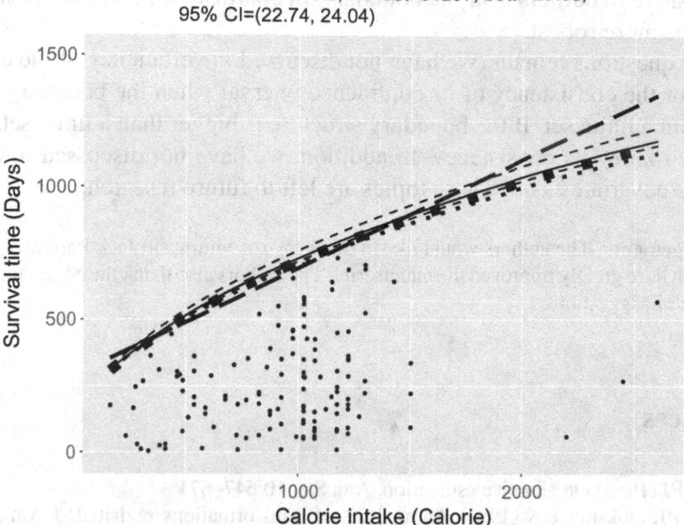

Fig. 4.4 The estimators after deleting the two observations with the longest and second-longest survival times. The notation is the same as that in Fig. 4.3

The consistent estimators $\hat{\beta}_{\hat{g}}$ and $\hat{\beta}_{\hat{g},PMLE}$ both show a positive trend as a function of calorie intake, whereas the two quantile regression estimators suggest a negative trend. To understand this phenomenon, we delete the two observations with the longest survival times (see Fig. 4.4). The best model is selected as $C = 0.5$, with $FitTest = 91297$ and $\hat{\beta}_{\hat{g}} = 23.08(22.74, 24.04)$. Now, all four estimators are very close and indicate a positive trend, suggesting that quantile regressions are very sensitive to outliers near the extreme quantile. On the other hand, the boundary support model estimates yield relatively similar curves, with or without the outliers.

4.7 Conclusion

In this paper, we have examined a semiparametric boundary support model with one covariate. A semiparametrically adaptive and efficient estimator is given in Theorem 4.2 when the boundary structure is known. The adaptive estimator and its confidence interval estimation based on limits of experiments framework were applied to a real-data analysis for the two-sample case.

Next, we proposed a selection method for the boundary structure when the structure with one continuous covariate is unknown. We also proved its consistency when the set of boundary structures is a finite set. The method was applied to investigate the relation between the maximum survival times of lung cancer patients and their calorie intake. Here, a quantile regression (quantile $\tau = 1$) cannot give a confidence region, and is sensitive to outliers near the boundary. In contrast, our consistent estimator is shown to be more robust.

Several questions remain. We have not discussed the efficiency of the consistent estimator or the consistency of its confidence interval when the boundary structure is not within a finite set. If the boundary structure is bigger than a finite set, then we do not even know its consistency. In addition, we have not discussed multivariate continuous covariate cases. These topics are left to future research.

Acknowledgements The authors would like to thank Ryozo Miura and Jack Porter for their comments, which have greatly improved the manuscript. The authors also thank the editor and reviewers for their useful suggestions.

References

1. Bickel PJ (1982) On adaptive estimation. Ann Stat 10:647–671
2. Bickel PJ, Doksum KA (1981) An analysis of transformations revisited. J Am Stat Assoc 76:296–311
3. Bickel PJ, Doksum KA (2015) Mathematical statistics: basic ideas and selected topics, Volume I. CRC Press, Boca Raton
4. Bickel PJ, Klaassen CA, Ritov Y, Wellner JA (1993, 1996) Efficient and adaptive estimation for semiparametric models. Springer, New York
5. Box GEP, Cox DR (1964) An analysis of transformations. J R Stat Soc 26:211–243
6. Chernozhukov V (2005) Extremal quantile regression. Ann Stat 33:806–839
7. Chernozhukov V, Hong H (2004) Likelihood estimation and inference in a class of nonregular econometric models. Econometrica 72:1445–1480
8. Donald SG, Paarsch HJ (2002) Superconsistent estimation and inference in structural econometric models using extreme order statistics. J Econ 109:305–340
9. Flinn C, Heckman J (1982) New methods for analyzing structural models of labor force dynamics. J Econ 18:115–168
10. Fraga Alves I, Neves C, Rosário P (2017) On extreme regression quantiles. Extremes 20:199–237
11. Ghosal S, Samanta T (1995) Asymptotic behavior of Bayes estimates and posterior distributions in multiparameter nonregular cases. Math Methods Stat 4:361–388
12. Hall P (1982) On estimating the endpoint of a distribution. Ann Stat 34:556–568
13. Hirano K, Porter JR (2003) Asymptotic efficiency in parametric structural models with parameter-dependent support. Econometrica 71:1307–1338
14. de Haan L, Ferreira A (2006) Extreme value theory. An introduction. Springer, Berlin
15. Ibragimov IA, Has' Minskii RZ (1981) Statistical estimation: asymptotic theory. Springer, New York
16. Knight K (2001) Limiting distributions of linear programming estimators. Extremes 4:87–103
17. Koenker R (2005) Quantile regression. Cambridge University Press, Cambridge
18. Le Cam L (1972) Limits of experiments. In: Proceedings of 6th Berkeley symposium on mathematical statistics and probability, vol 1. University of California Press, Berkeley, California, pp 245–261
19. Loprinzi CL, Laurie JA, Wieand HS, Krook JE, Novotny PJ, Kugler JW, Bartel J, Law M, Bateman M, Klatt NE (1994) Prospective evaluation of prognostic variables from patient-completed questionnaires. North Central Cancer Treatment Group. J Clin Oncl 12:601–607
20. Ozeki A (2012) Efficient inference in semiparametric models. Updated printing. Dissertation, University of Wisconsin Madison

21. Portnoy S, Jurečková J (1999) On extreme regression quantiles. Extremes 2:227–243
22. Smith RL (1985) Maximum likelihood estimation in a class of nonregular cases. Biometrika 72:67–90
23. Smith RL (1987) Estimating tails of probability distributions. Ann Stat 15:1174–1207
24. Smith RL (1994) Nonregular regression. Biometrika 81:173–183
25. van der Vaart AW (1991) An asymptotic representation theorem. Int Stat Rev 59:97–121
26. van der Vaart AW (1998) Asymptotic statistics. Cambridge University Press, Cambridge
27. Yu P (2015) Adaptive estimation of the threshold point in threshold regression. J Econ 189:83–100

Chapter 5
Comparison of AMS and POT Analysis with Long Historical Precipitation and Future Change Analysis Using "d4PDF"

Shigenobu Tanaka

Abstract This paper assesses extreme precipitation events, which are one of the most impactful hydrological circulation events for policy decisions. In extreme analysis, samples of block maxima or peaks over threshold (POT) are used. However, annual maximum series (AMS), a type of block maxima, has been employed more often than POT. This study deals with problems that often occur in extreme analysis with long historical records and large ensembles of climate simulations. The following work is carried out: (1) AMS analysis is compared with POT analysis using long-term historical precipitation records at meteorological stations in Tokyo and Nagoya. With a carefully selected threshold, the generalized Pareto distribution keeps a more stable shape parameter than the generalized extreme value (GEV) distribution for AMS and gives relatively reliable return levels along with accumulated observation; and (2) a method using 60-year maxima is introduced to manage a very large set of AMS samples to which both the Gumbel distribution and the GEV distribution cannot fit well. Figures to obtain the 100-year return level are prepared based on the Gumbel distribution through the examination of relationships among past and future statistics of 1- to 3-day precipitation with a large ensemble of climate simulations called d4PDF. It is interesting that there are common statistical characteristics among the 1-day, 2-day, and 3-day precipitations.

Keywords AMS · Climate change · d4PDF · Shape parameter · Threshold · POT · Return level

S. Tanaka (✉)
DPRI, Kyoto University, Gokasho, Uji 611-0011, Japan
e-mail: tanaka.shigenobu.4m@kyoto-u.ac.jp

N. Hoshino et al. (eds.), *Pioneering Works on Extreme Value Theory*,
JSS Research Series in Statistics,
https://doi.org/10.1007/978-981-16-0768-4_5

5.1 Introduction

Let us begin with a quotation from UNESCO [13]:

> Hydrology is the science which deals with the waters of the earth, their occurrence, circulation and distribution on the planet, their physical and chemical properties and their interactions with the physical and biological environment, including their responses to human activity. Hydrology is a field which covers the entire history of the cycle of water on the earth.

Precipitation is one of the most important components of hydrological circulation and is directly connected to surface runoff. The effective duration of precipitation depends on the spatial characteristics of the drainage area. The total amount of precipitation in the effective duration is generally used to design flood control structures for the target drainage area. In Japanese river basins, effective durations range from several hours to three days. The design of flood control plans reflects historical data extremes. The period from the planning until the completion of flood control structures presently requires a longer time than in the past. The assessment of future precipitation increases and their impact on basins is urgently required.

In many floods, severe damage results not only from high water levels and flooding, but also from large-scale debris flows and sedimentation, impacting the lives of the population long after the hazards are gone. The Intergovernmental Panel on Climate Change Assessment Report 5 [3] warned that, in the future, "extreme precipitation events over most of the mid-latitude land masses and wet tropical regions will very likely become more intense and more frequent." In many reports on climate change monitoring [5] and the risk of occurrence of very large disasters [8], the same concerns have been raised and confirmed.

Flood and debris-flow disasters are generally caused by extreme rainfall. Flood risk management requires an adequate understanding of relationships between hazardous events and their frequency.

Historically, observational data have been used for extreme analysis. However, because observation periods are generally short, additional data tend to affect return levels, based on which the design rainfall of a flood control plan is determined. From the perspective of infrastructure construction, a high priority should be placed on the stability of return levels so that these levels will not be largely affected by additional data.

In this study, the performance of extreme distributions is examined with long-term precipitation observation records. Additionally, the impact of climate change on extreme precipitation events is studied using a large ensemble of climate simulations with a 20 km regional climate model focusing on six major river basins.

5.2 Data Used

In this study, two kinds of precipitation data are analyzed; one is long daily precipitation at Tokyo and Nagoya, and the other is a large ensemble of climate simulations for both past and future climate conditions.

5.2.1 Historical Precipitation Observations

To examine the performance of the extrapolated return levels, we employ daily precipitation data observed at a meteorological station in Tokyo and another in Nagoya. Among the oldest meteorological stations in Japan, these two stations have observed daily precipitation since 05 June 1875 and 01 July 1890, respectively. Figure 5.1 shows the time series of annual maximum daily precipitation at these stations. The records of the first year were excluded, because of the short periods of observation.

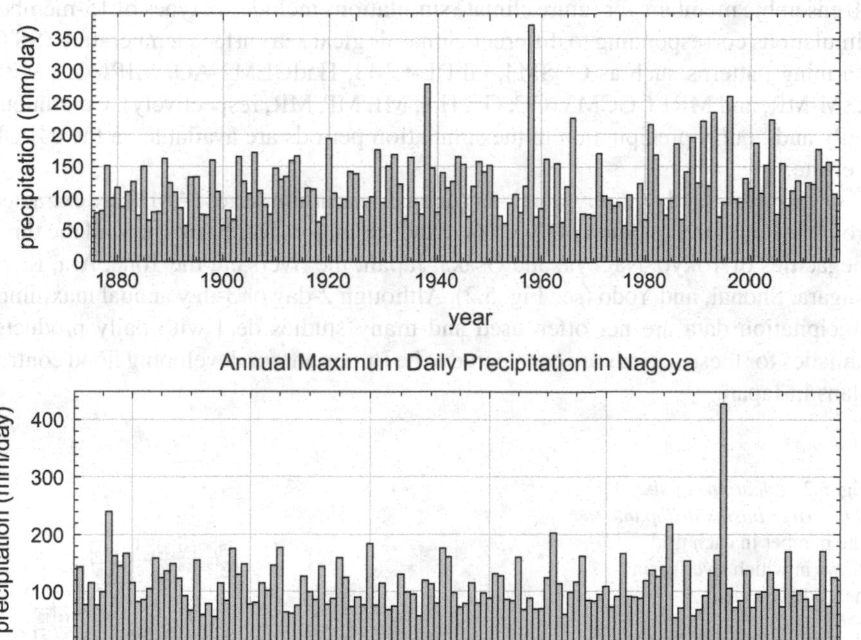

Fig. 5.1 Time series of annual maximum daily precipitation at a meteorological station in Tokyo (top) and one in Nagoya (bottom), Japan

The Mann–Kendall test for both sites shows that the time series do not change with time.

In both sites, the level of daily rainfall is not very remarkable, except for historical high records. Tokyo has one rainfall event of over 200 mm/day before 1958, and Nagoya has two rainfall events of over 200 mm/day before 2000. The historical maximum record is 371.9 mm/day for Tokyo and 428 mm/day for Nagoya.

5.2.2 Large Ensemble of Climate Simulations

A large ensemble of climate simulations with a 60 km atmospheric general circulation model and dynamical downscaling with a 20 km regional climate model (RCM) were prepared to obtain probabilistic future projections of low-frequency local-scale rainfall events [9]. The simulation outputs are available in the "Database for Policy Decision-Making for Future Climate Change" (d4PDF), which is intended to be utilized for impact assessment studies and adaptation planning for global warming. The RCM simulations consist of historical climate simulations (1951–2010, 50 ensemble members) and future climate simulations (2051–2110, 90 ensemble members). The 90 ensemble members for future climate simulations include six types of 15-member simulations corresponding to different climatological sea surface temperature (SST) warming patterns such as CCSM4, GFDL-CM3, HadGEM2-AO, MIROC5, MPI-ESM-MR, and MRI-CGCM3 (CC, GF, HA, MI, MP, MR, respectively). Continuous daily and hourly precipitation in the simulation periods are available on the d4PDF website.

In this paper, 1-day, 2-day, and 3-day annual maximum series (AMS) are extracted from the daily precipitation of 99 grids in the six major river basins around the three megacities of Tokyo, Nagoya, and Osaka, Japan; the rivers are the Tone, Ara, Kiso, Nagara, Shonai, and Yodo (see Fig. 5.2). Although 2-day or 3-day annual maximum precipitation data are not often used and many studies deal with daily products, statistics for these periods are considered to be important for developing flood control plans in Japan.

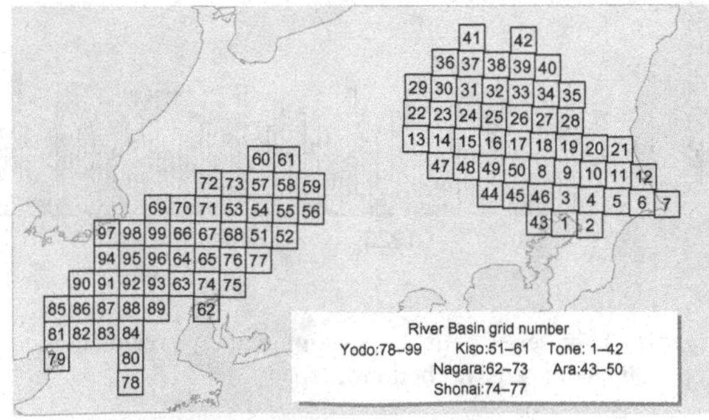

Fig. 5.2 *Locations of six major river basins in Japan.* The number in each grid shows in which river basin the grid is located

5.3 Frequency Analysis

Extreme value distributions are applied to estimate very infrequent quantiles [1, 12]. The generalized extreme value distribution (GEV) is used for block maxima while the generalized Pareto distribution (GP) is used for peaks over threshold (POT). Both GEV and GP are three-parameter distributions; however, when the shape parameters of GEV and GP are zero, they become two-parameter distributions (see Table 5.1). GEV becomes the Gumbel distribution (Gumbel) and GP the exponential distribution (Exp). When dealing with block maxima, the simplest distribution is the Gumbel distribution, which has two parameters closely related to block maxima statistics, namely the location parameter ξ and the scale parameter α; these parameters have the following relationship with the mean μ and standard deviation σ:

$$\alpha = \sqrt{6}\sigma/\pi, \quad \xi = \mu - \gamma\alpha, \tag{5.1}$$

where γ is the Euler–Mascheroni constant; $\gamma = 0.5772\cdots$. To estimate α and ξ, μ and σ of (5.1) are replaced with the sample mean $\bar{x}$ and standard deviation s, respectively. The corresponding estimates are denoted by $\hat{\alpha}$ and $\hat{\xi}$. One can estimate the non-exceedance probability F and a return period with cumulative distribution function, or a return level using Eq. (5.2) or (5.3)

$$x_F = \hat{\xi} - \hat{\alpha}\log(-\log(F)). \tag{5.2}$$

$$x_F = A\bar{x}, \quad A = 1 - (\gamma + \log(-\log(F)))\frac{\sqrt{6}}{\pi}\frac{s}{\bar{x}}. \tag{5.3}$$

For independent events, $F(x)$ for AMS analysis is connected with $G(x)$ for POT analysis by the Poisson distribution

$$F(x) = \exp\left(-\lambda(1 - G(x))\right), \tag{5.4}$$

where λ is the arrival rate, which is equal to the average number of events per year larger than a threshold value [12].

Table 5.1 *The cumulative distribution functions of extreme value distributions*

Data	Three parameters distribution	Two parameters distribution
AMS	GEV: $F(x) = \exp\left(-\left(1 - \kappa\frac{x-\xi}{\alpha}\right)^{1/\kappa}\right)$	Gumbel: $F(x) = \exp\left(-\exp\left\{-\frac{x-\xi}{\alpha}\right\}\right)$
POT	GP: $G(x) = 1 - \left(1 - \kappa\frac{x-\xi}{\alpha}\right)^{1/\kappa}$	Exp: $G(x) = 1 - \exp\left\{-\frac{x-\xi}{\alpha}\right\}$

(ξ :location parameter, α:scale parameter, κ:shape parameter)

5.3.1 Block Size of Block Maxima

AMS has been conventionally used as a kind of block maxima, and its block size is one year. However, in the case of large data ensembles, the selection of a block size requires careful considerations. The d4PDF is a large database that has 50 or 15 ensemble members of 60-year simulations. As a serial simulation, this will be converted to 3,000-year AMS for the historical climate simulation and 900-year AMS for the future simulation. On the other hand, if the 60-year period is considered a block, there are 50 samples of "60-year maxima." If a sample comes from a common Gumbel distribution, the points will be placed on an almost straight line on the Gumbel probability paper. Different block sizes show different locations, however, they have the same scale parameter, which corresponds to the slope of the plots. We observe that slopes are very similar between the 100-year return period of the AMS plot and the 60-year maxima plot in Fig. 5.3. In the probability plot of AMS, the lower part is different from the upper part. Regarding the 100-year return period, the plotted points show a return level that disagrees with those of the Gumbel and GEV distributions. In the AMS plots, precipitation of less than 90 mm occupies half of the sample and shows a different tendency from that of the upper part, which might be a reason for the disagreement. On the other hand, the minimum value of the 60-year maxima is 191.2 mm, and the 60-year maxima series does not include very small values of less than 90 mm. Both the Gumbel and GEV distributions adequately fit the probability plot of the 60-year maxima. To compare the 100-year return level of AMS with the 60-year maxima, different non-exceedance probabilities are needed; $F = 0.99$ for AMS and $F = 0.99^{60} = 0.5472$ for the 60-year maxima. Akaike's Information Criterion (AIC) recommends the GEV distribution for AMS and the Gumbel distribution for the 60-year maxima. Considering the 100-year return level, there is a discrepancy between the AMS probability plots and the fitted GEV distribution and an agreement between the probability plots of the 60-year maxima and both distributions. For all of the 99 grids, the probability plots of the 60-year maxima and the Gumbel distribution are in good agreement at $F = 0.5472$. Based on these observations, in this study, the 60-year maxima are used to estimate the 100-year return level. However, AMS is essential in practical situations, and consequently, in this study, the relationship between AMS and the 60-year maxima is prepared.

5.3.2 Threshold Selection

The threshold selection is crucial in POT analysis because POT samples depend heavily on a threshold. The sample mean excess function (SMEF) is a traditional tool for selecting a threshold. The threshold selection method using SMEF is based on the fact that, if the excesses of a threshold follow the Exp distribution, the mean of the excesses is equal to the scale parameter [10]. In practice, a threshold candidate range where SMEF is constant is determined, and the smallest point of the range is

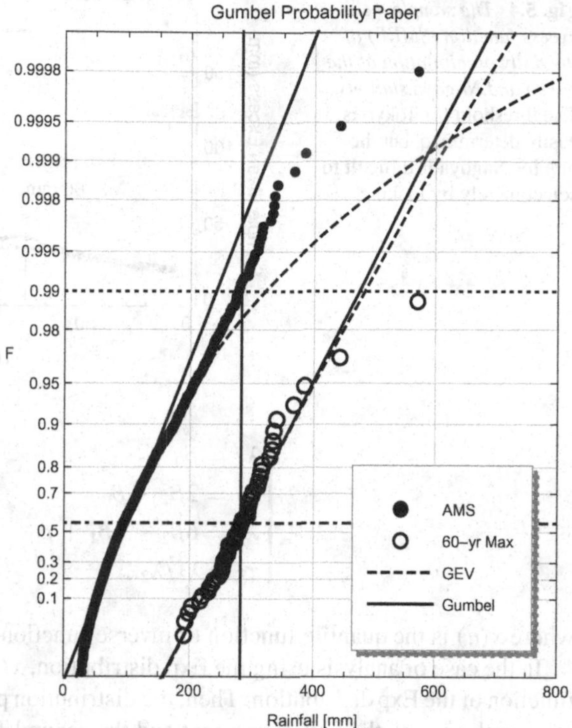

Fig. 5.3 *A comparison of the probability plots and fits of annual maximum series (AMS) with the 60-year maxima of grid 47 of the Ara river basin in Fig.* 5.2. The dotted line shows the 100-year return period for AMS and the dot-dashed line shows the 100-year return period for the 60-year maxima. The vertical solid line indicates the 100-year return level of the Gumbel distribution for the 60-year maxima

selected. However, this method is subjective. Figure 5.4 shows an example of the SMEF graph for Tokyo and Nagoya. The threshold is evident in the case of Tokyo, but is unclear in the case of Nagoya. This shows the importance of setting an appropriate threshold so that the extracted POT data can be adequately fitted with the GP or Exp distribution. Thus, the SMEF method to determine a candidate threshold for the Exp distribution can potentially lead the shape parameter of the GP distribution to be around 0, which is mostly not appropriate. Therefore, in this study, the author proposes an automatic objective method using the L-moment relationship to address this issue.

The maximum likelihood estimation is generally used for parameter estimation in extreme analysis; however, the L-moment method is less affected by outliers [2]. When probability weighted moments are expressed by Eq. (5.5), L-moments λ_i are expressed by Eq. (5.6)

$$\beta_r = \int_0^1 x(u)u^r \, du \,. \tag{5.5}$$

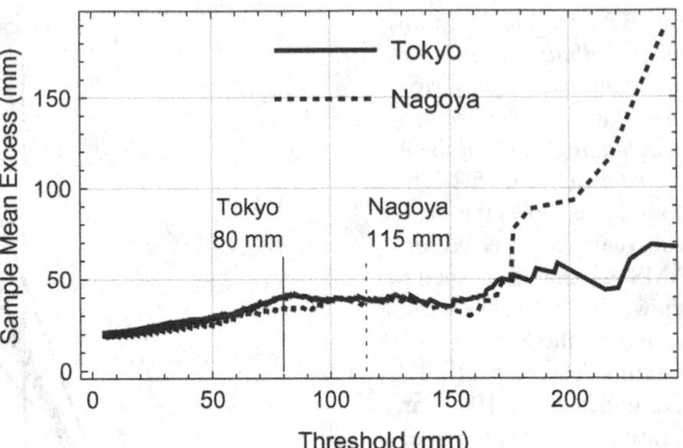

Fig. 5.4 *The sample mean excess function (SMEF) of the daily precipitation at the Tokyo and Nagoya stations.* The threshold for Tokyo is easily determined, but the one for Nagoya is difficult to select merely by looking

$$\begin{cases} \lambda_1 = \beta_0 \\ \lambda_2 = 2\beta_1 - \beta_0 \\ \lambda_3 = 6\beta_2 - 6\beta_1 + \beta_0 \\ \tau_3 = \lambda_3/\lambda_2 \end{cases} \tag{5.6}$$

where $x(u)$ is the quantile function or inverse function of the distribution function.

In the case of analysis using the Exp distribution, $x(u)$ is replaced by the quantile function of the Exp distribution. Then, the distribution parameters have the following relationships with the first L-moment and the second L-moment:

$$\begin{cases} \lambda_1 = \xi + \alpha \\ \lambda_2 = \alpha/2 \end{cases} \tag{5.7}$$

where ξ is the location parameter and α is the scale parameter. Equations (5.7) yield Eq. (5.8) as a fraction of them. When Eq. (5.8) is equal to one, the scale parameter value derived from the mean of the excesses of the threshold candidate is equal to that from the variance of the excesses of the threshold candidate, and the independent peaks over the threshold can be fitted by the Exp distribution. Thus, Eq. (5.8) can be considered as a "fidelity index" for the Exp distribution. If there are several threshold candidates that make Eq. (5.8) equal to one, the smallest candidate should be selected to obtain a larger sample size.

$$\text{Fidelity index for Exp:} \quad (\lambda_1 - \xi)/(2\lambda_2)\,. \tag{5.8}$$

Similarly, from the relationships among the first, second, and third L-moments and the parameters of the GP distribution shown in Eq. (5.9), Eq. (5.10) is given as an index that shows fidelity for the GP distribution.

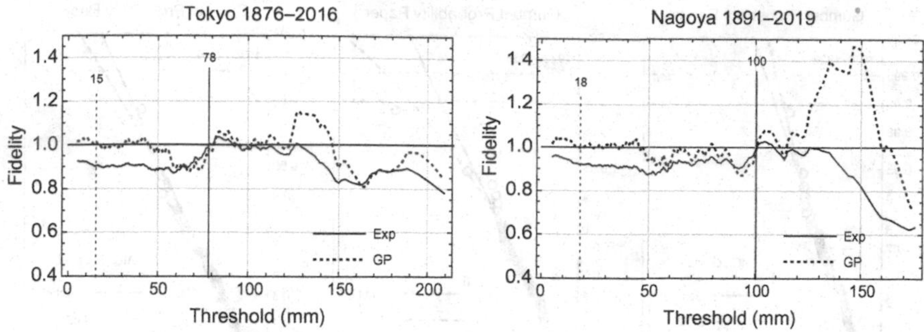

Fig. 5.5 *Examples of the proposed threshold selection method for Tokyo (left) and Nagoya (right)*

$$\begin{cases} \lambda_1 = \xi + \alpha/(1+\kappa) \\ \lambda_2 = \alpha/((2+\kappa)(2+\kappa)) \\ \tau_3 = (1-\kappa)/(3+\kappa) \end{cases} \tag{5.9}$$

$$\text{Fidelity index for GP:} \quad \frac{(\lambda_1 - \xi)/\lambda_2 - 1}{2(1-\tau_3)/(1+\tau_3)}. \tag{5.10}$$

Figure 5.5 shows the detected thresholds for the Exp and GP distributions for Tokyo and Nagoya. For Tokyo, the threshold is 78 mm for Exp and 15 mm for GP. GP may also fit POT data over 78 mm, but with a shape parameter very close to 0; however, Exp no longer fits at a threshold smaller than 78 mm. For Nagoya, the threshold is 100 mm for Exp and 18 mm for GP.

5.3.3 AMS and POT Analysis of Long Historical Precipitation

To examine the performance of the AMS and POT analysis for estimating extrapolated return levels, daily precipitation data from Tokyo and Nagoya were used.

First, the AMS analysis was conducted using 141 years of data from Tokyo. The whole AMS record was divided into three periods with the same length, and the AMS of each period was plotted and fitted with the Gumbel and GEV distributions. Figure 5.6 plots the three data periods with AIC recommendations. The GEV distribution has an upper bound for the first period; it has a very thick tail with a very large record for the second period, in which the maximum historical record was observed, but the tail becomes thinner in the third period. The GEV and Gumbel distributions are preferred by AIC for the second and third periods, respectively.

Second, the block size was changed. The maximum values were extracted with different block sizes from the whole record. Figure 5.7 shows a comparison of the plots of the AMS, 2-year maxima, and 3-year maxima. The Gumbel and GEV distri-

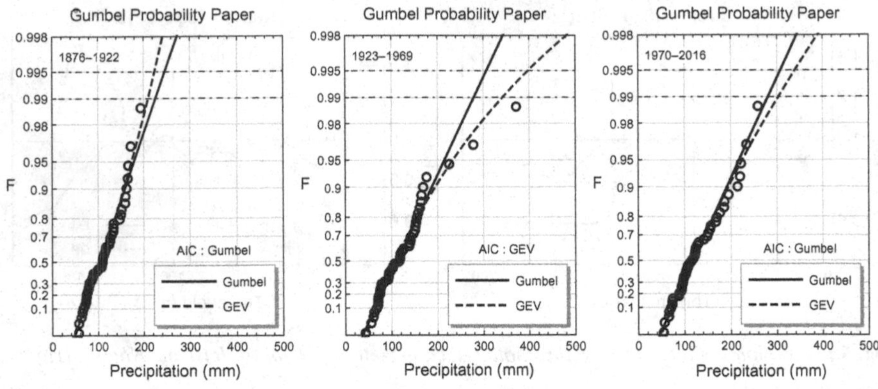

Fig. 5.6 *Probability plots of the three divided periods with fits for the Tokyo data.* AIC : distribution recommended using Akaike's Information Criterion

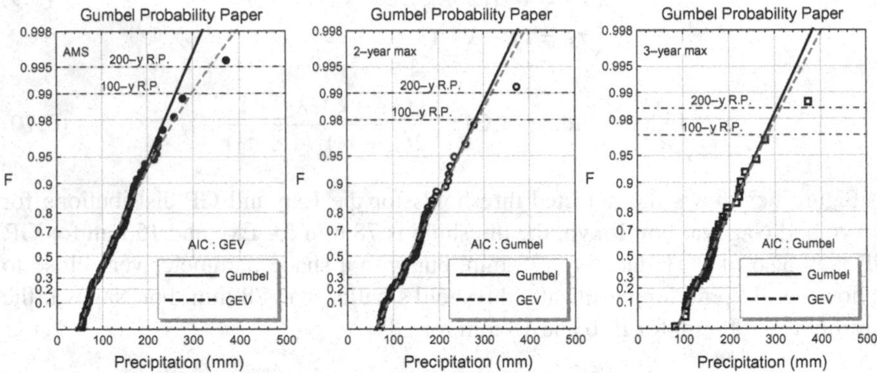

Fig. 5.7 *Probability plots of m-year maxima with fits for Tokyo*

butions differ by a large amount at large return levels with AMS, but the discrepancy becomes small with the 2- and 3-year maxima. One-quarter of the AMS samples are less than 80 mm and their plots have different tendencies from the others. These relatively small data show a smaller return level of the Gumbel distribution and a larger return level of the GEV distribution for a longer return period. In Fig. 5.7, the GEV distribution is recommended by AIC for the AMS case, but the Gumbel distribution is recommended for the 2-year and 3-year maxima. Regarding the 2-year and 3-year maxima, small-value data in AMS have disappeared, and the GEV distribution looks very similar to the Gumbel distribution. However, neither the Gumbel distribution nor the GEV distribution fits the outliers. This situation is different from Sect. 5.3.1 and requires further examination.

Third, both AMS and POT analyses were applied. Figure 5.8 shows the evolution of the 200-year return level of the Gumbel, GEV, Exp, and GP distributions along with data accumulation (top graph). Daily precipitation has been recorded at the

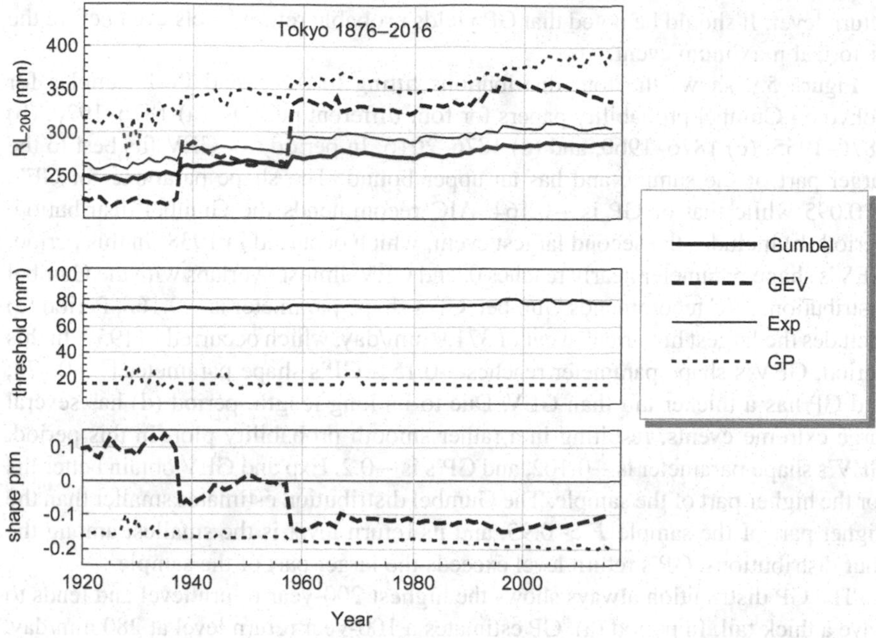

Fig. 5.8 *The evolution during 1920–2016 of the 200-year return level (top), thresholds of POT (middle), and shape parameters of GEV and GP (bottom), along with data accumulation from 1876 at Tokyo*

Tokyo meteorological station since 1876; 200-year return levels for the observation period from 1876 to any year (which appears in the horizontal axis) are plotted in Fig. 5.8. The Gumbel and GEV distributions were applied to AMS data while the Exp and GP distributions were applied to POT data. To extract the POT data, the thresholds for both Exp and GP distributions were selected for each set of samples using the method introduced in the previous section (middle graph of Fig. 5.8). There are small variations in the threshold for both the Exp and GP distributions while the threshold of GP is almost constant after 1970. The bottom graph shows the evolution of the shape parameter of the GEV and GP distributions. The GEV's shape parameter changes from 0.1 to −0.1, and sharp drops appear at every critical event in 1938 and 1958. On the other hand, the GP's shape parameter is always negative and settles at around −0.2 after 2000. It may be possible to say that Exp estimates the most stable return level, and Gumbel estimates the second-most stable return level. However, their return levels are moderate, and this trend does not change, even after the occurrence of the historical maximum event in 1958. The return level of GEV changes in response to the large rainfall event, but it is too small before 1938. With the effect of the large rainfall events in 1938, GEV's return level becomes very similar to those of Gumbel and Exp; however, it drops again to the lowest among the four distributions just before the largest event in 1958. GP always yields the highest return level. It should be noted that GP yields probable return levels even before the historical maximum event.

Figure 5.9 shows the four distributions fitting to AMS and POT samples for Tokyo on Gumbel probability papers for four different periods: (a) 1876–1920, (b) 1876–1955, (c) 1876–1960, and (d) 1876–2016. In period (a), GEV fits best to the larger part of the sample and has an upper bound. The shape parameter of GEV is 0.095 while that of GP is −0.164. AIC recommends the Gumbel distribution. Period (b) includes the second largest event, which occurred in 1938. In this period, GEV's shape parameter nearly reaches 0, and GEV almost overlaps with the Gumbel distribution. AIC recommends Gumbel. GP's shape parameter is −0.166. Period (c) includes the largest historical event of 371.9 mm/day, which occurred in 1957. In this period, GEV's shape parameter reaches −0.151. GP's shape parameter is −0.178, and GP has a thicker tail than GEV. Due to its long length, period (d) has several large extreme events, resulting in a rather smooth probability plot. In this period, GEV's shape parameter is −0.102, and GP's is −0.2. Exp and GEV obtain better fits for the higher part of the sample. The Gumbel distribution estimates smaller than the higher part of the sample $F > 0.95$, and its return level is the smallest among the four distributions. GP's return level exceeds the larger part of the sample.

The GP distribution always shows the highest 200-year return level and tends to have a thick tail. In period (a), GP estimates a 100-year return level at 280 mm/day, although the maximum in the period is less than 200 mm/day. However, the second largest event, with a rainfall of 278.3 mm/day, occurred in 1938. In periods of (a) and (b), only GP estimates a 200-year return level of over 300 mm/day, although a rainfall event of 371.9 mm/day occurred in 1958. In period (c), GP obtains a very close fit to the sample plots. In period (d), GP estimates a much higher return level than the sample. At present, rainfall of more than 400 mm/day is not rare in Japan. In this research, a threshold of 80 mm/day was selected by SMEF with the data of the whole period (see Fig. 5.4). When it is used for different periods, GP estimates 200-year return levels at around 300 mm/day, which is the largest among the four distributions before 1970 and the second largest after 1970. From this analysis, it is clear that the threshold should be obtained separately for each sample and distribution.

Although the sample does not include large extremes and GEV's shape parameter is positive, as seen from the top left graph in Fig. 5.9, GP estimates a high return level. Figure 5.10 shows the performance of GP for Nagoya, comparable to the same analysis for Tokyo (Fig. 5.8). The top figure shows that only GP constantly estimates a rainfall of more than 300 mm/day for a 200-year return level, even before 2000, in which a record-breaking event of 428 mm/day occurred. The threshold of GP's is more stable than that of Exp, and the former's shape parameter maintains at around −0.2. (middle and bottom). The return levels of GEV, Gumbel, and Exp almost coincide, however, only GP shows more than 300 mm in 1935 (top).

The author has checked the performance of the proposed threshold selection method for rainfalls of 1 to 96 hours for 17 meteorological stations in Japan, whose observation periods are longer than 80 years. The results show that GP's thresholds are very small in most cases and that the shape parameters stay at around −0.2.

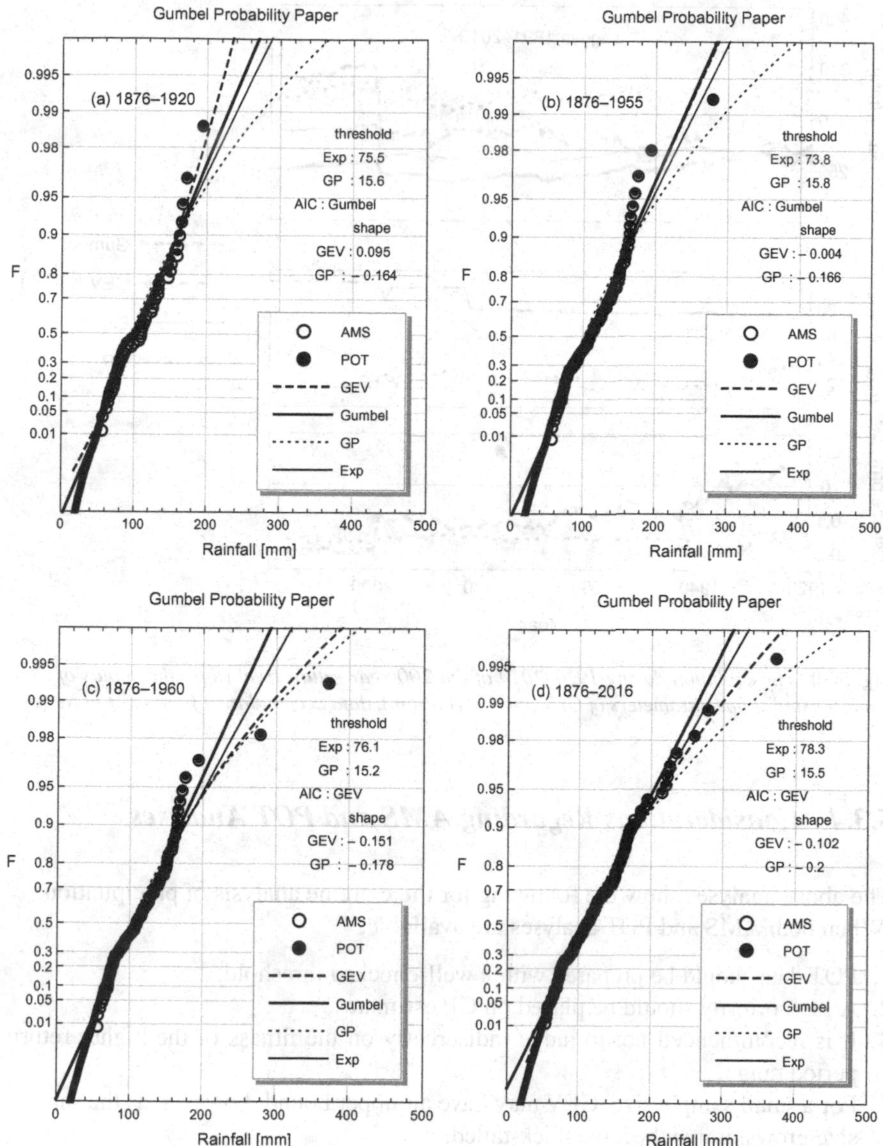

Fig. 5.9 *Four distributions fitting to AMS and POT samples for Tokyo on Gumbel probability papers for four different periods*: **a** *1876–1920*, **b** *1876–1955*, **c** *1876–1960, and* **d** *1876–2016*

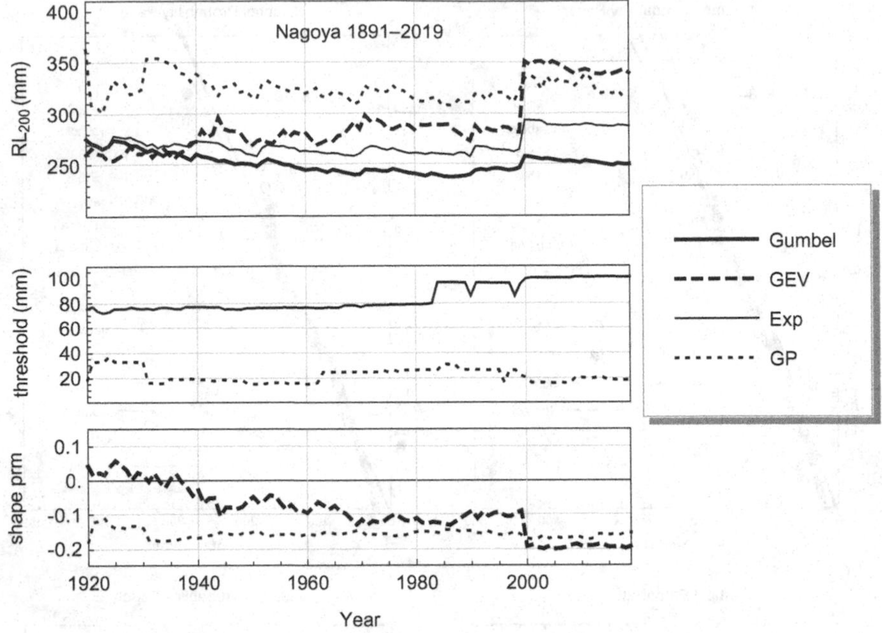

Fig. 5.10 *The evolution during 1920–2019 of the 200-year return level (top), thresholds of POT (middle), and shape parameters of GEV and GP (bottom), data accumulation from 1891 at Nagoya*

5.3.4 Considerations Regarding AMS and POT Analyses

The above analyses show the following for the extreme analysis of precipitation: When both AMS and POT analyses are available,

1. POT data should be prepared with a well-checked threshold.
2. A high priority should be placed on GP estimates.
3. It is recommended not to judge indiscreetly on the fitness of the higher return period range.
4. For a small sample size, GEV may have an upper bound; however, as the sample size grows, it may become thick-tailed.
5. For a small sample size, while the Gumbel distribution may estimate a higher return level than the sample plot, its higher return level often becomes lower as the sample size grows. Thus, the return level estimate of the Gumbel distribution should be taken as a lower reference value.
6. The performance of Exp depends on the threshold. Return levels obtained using the Exp distribution are slightly larger than those of the Gumbel distribution mostly in the examples of Sect. 5.3.3 (Figs. 5.8, 5.9 and 5.10).

When only AMS analysis is available,

1. According to AIC, GEV or Gumbel should be selected.
2. For a small sample size, when GEV has an upper bound, Gumbel should be selected.
3. When the probability plot is almost linear around the target return period but not well fitted with both GEV and Gumbel, n-year maxima may be able to fix the discrepancy, though the discrepancy with outliers may not be fixed. In this scenario, Gumbel and GEV are no longer useful, and it is better to make judgments based on the sample probability plot.

5.4 Analysis of Climate Simulations

Disaster countermeasures, such as flood control structures, are generally planned to prevent frequent disasters based on past events. The impact of climate change on the probability of extreme precipitation exceeding the design precipitation for disaster countermeasures raises the question of their capacity against flood risks. If river water overflows an embankment, the main structures, which are usually constructed from soil, will be at higher risk of breach. Therefore, it is crucially important to calculate the occurrence probability of extreme precipitation exceeding the design level for flood control plans.

Large ensemble climate simulations, such as the d4PDF, are used to assess the impact of climate change on extreme precipitation. The d4PDF is a database for policy decision-making to cope with future climate change and allows users to explicitly investigate the probability density function of extreme events. However, if large ensemble climate simulations are regarded as continuous serial data of 3000 years or 5400 years, the Gumbel distribution forms a straight line, as shown in Sect. 5.3.1. The results of this research were analyzed using the Gumbel distribution with the 60-year maxima (as stated in Sect. 5.3.1) since it is common to use a 100-year return level when developing flood control plans in Japan.

Figure 5.11 shows the relationship between the mean of the 60-year maxima and the 100-year return level for 1-day, 2-day, and 3-day precipitation in 99 grids in the six major river basins in Japan for the past climate. The slope of the regression line is 0.988, which is equal to A in Eq. (5.3). The reason why A is so close to 1 is that the second term of A is - 0.056 $s/\bar{x}$ with $s/\bar{x}$ being around 0.2–0.3 in the past climate and 0.1–0.4 in the future climate. Thus, the 100-year return level for 1-day, 2-day, and 3-day precipitation is considered to be roughly the same as the ensemble mean of the 60-year maxima.

Figure 5.12 shows a comparison of the ensemble mean of the 60-year maxima of 1-day, 2-day, and 3-day precipitation in the past climate and the future climate for each of the six river basins. This figure also shows the average future change ratio of each basin. They range from 1.16 to 1.32 and the average slope of all the 99 grids is 1.26. The range of the variation of the future 60-year maxima in the 99 grids is 1 to 1.7 times that of the past 60-year maxima except for the SST warming pattern MI,

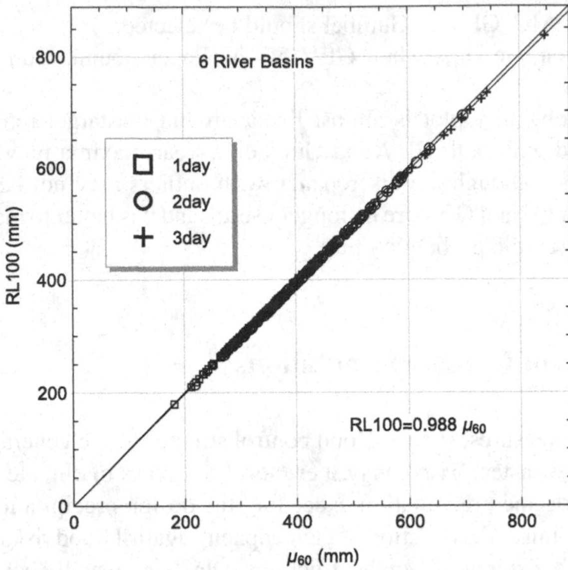

Fig. 5.11 *The relationships between the 100-year return level and the ensemble mean of the 60-year maxima for 1-, 2-, and 3-day precipitation for each of the 99 grids in the six river basins in Japan under the past climate*

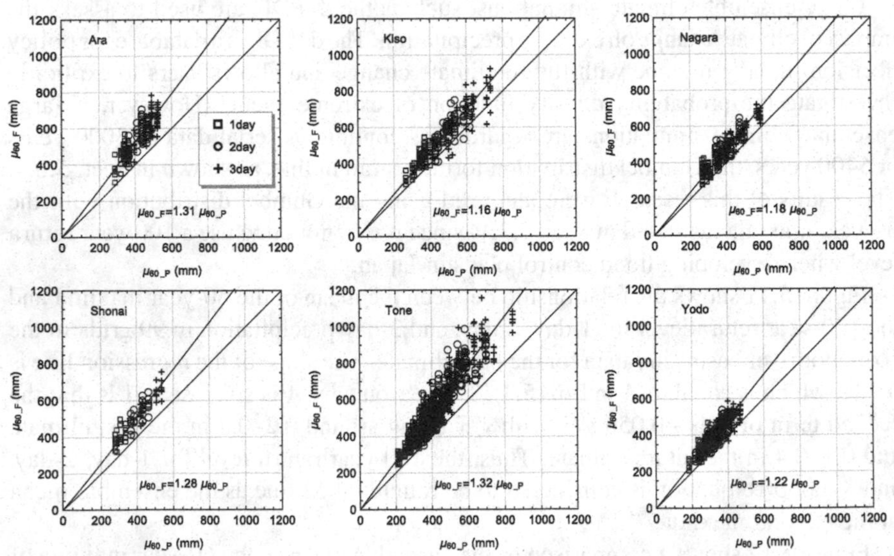

Fig. 5.12 *Comparisons of the ensemble means of the 60-year maxima in the past climate and the future climate*

whose range is 0.87–1.84. Only the Kiso River basin includes decreasing grids, and the Tone River basin includes extremely increasing grids in the MI SST.

Finally, the relationship between the ensemble mean of the 60-year maxima and the mean of AMS was analyzed. Figure 5.13 shows the relationship between the 1-day, 2-day, and 3-day precipitation in the past climate for each river basin. The ratio is slightly different from basin to basin, ranging from 2.15 to 2.48. Interestingly, the regression lines of the 1-day, 2-day, and 3-day precipitation are almost the same for each river basin. In the Kiso and Tone River basins, there are several grids where the ensemble mean of the 60-year maxima is greater than 600 mm: Four grids in the Kiso River basin and eight grids in the Tone River basin for the 3-day precipitation and one grid in the Kiso River basin and two grids in the Tone River basin for the

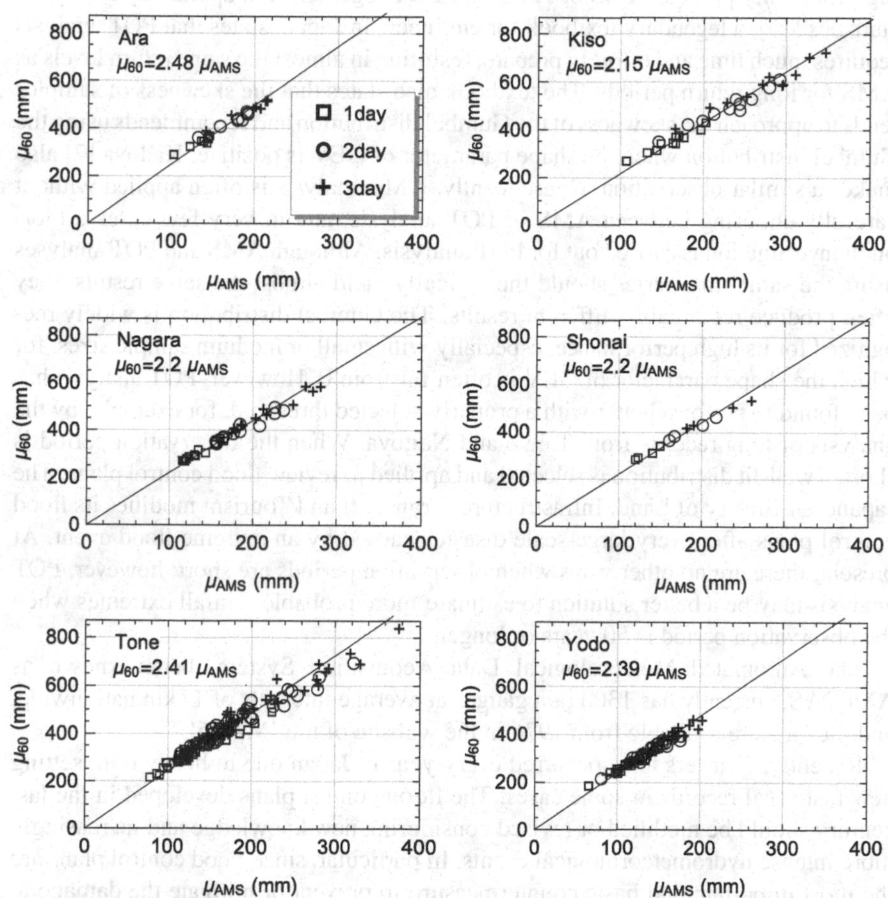

Fig. 5.13 *Comparisons of the mean of AMS and the ensemble mean of the 60-year maxima in the past climate*

2-day precipitation. These very large precipitation values will be multiplied in the future climate.

More detailed information is needed in order to analyze how much precipitation will fall in the drainage basin upstream of the flood control reference point, in which the precipitation within the duration of rainfall, which is used to develop flood control plans, is strongly related to the peak flood discharge.

5.5 Discussion

The original motivation for this study came from the author's experience of drawing probability plots of AMS or AMS and POT together. "An applied hydrological statistics" [4], a legendary textbook for engineers in Japan, states that POT analysis requires much time and effort to prepare, resulting in almost the same return levels as AMS for long return periods. The textbook also states that the skewness of samples tends to approach the skewness of the Gumbel distribution and recommends using the Gumbel distribution when the shape parameter of GEV is positive. Kadoya [7] also makes a similar observation. Consequently, AMS analysis is often applied without carefully choosing between AMS or POT analysis, and in very few cases a thorough investigation is carried out for POT analysis. Although AMS and POT analyses using the same time series should theoretically yield almost the same results, they often produce remarkably different results. The Gumbel distribution is widely recognized for its high performance, especially with small or medium sample sizes, for which the shape parameter of GEV is often far from 0. However, POT analysis has been found to perform better with a properly selected threshold, for example, by the analysis of long records from Tokyo and Nagoya. When the observation period is short, a well-fit distribution is selected and applied to review flood control plans. The Japanese Ministry of Land, Infrastructure, Transport and Tourism modifies its flood control plans after every large-scale disaster caused by an extreme flood event. At present, there are no other ways when observation periods are short; however, POT analysis may be a better solution to estimate more probable rainfall extremes when the observation period is 50 years or longer.

The Automated Meteorological Data Acquisition System, better known as AMeDAS, currently has 1300 rain gauges at average intervals of 17 km nationwide, and the data are available from 1976 in the website of the JMA [6].

Recently, disasters have occurred every year in Japan due to heavy rain, setting new historical records in some cases. The flood control plans developed in the last century should be modified or revised considering new knowledge and increasingly more intense hydrometeorological events. In particular, since flood control plans are the most important and basic countermeasure to prevent or mitigate the damage to urban areas from rainfall, hydrological extreme analysis is more important than ever.

In order to use d4PDF in flood control planning, it is necessary to carry out a more detailed examination and bias correction using observed data. Further investigation of bias correction is required; however, as shown in Figs. 5.12 and 5.13, the common statistical characteristics among 1-day, 2-day, and 3-day precipitation may be useful information for checking biases.

5.6 Conclusion

This research presents findings on extreme analysis using long-term observational data. When data are available only for a short period, the analysis needs to rely on a goodness-of-fit index, such as AIC. However, with long accumulated records, this research has found that a goodness-of-fit index does not necessarily perform well for extrapolation, possibly due to smaller extremes, outliers, or both. For such cases, the GP distribution with POT exceeding a carefully selected threshold may give superior results compared to other distributions. When only AMS is available, either the GEV or Gumbel distributions should be selected according to AIC. The Gumbel distribution with m-year maxima is a possible alternative to the GP distribution; however, if outliers exist with m-year maxima, the Gumbel and GEV distributions are no longer useful, and it is better to judge based on sample probability plots.

Many researchers in the field of climate change impact assessment have been tackling bias correction [11, 14]. In the past, a quantile mapping correction method, which compares the quantile of a set of observation with that of a simulation, was widely applied for correcting biases. However, with the advent of very large ensemble databases such as the d4PDF, the conditions of bias recognition have significantly changed. No one knows which ensemble product among large ensemble members should be compared with observation. The ensemble mean also cannot be appropriate for comparison. This situation looks as if this field has lost its compass for the truth, unable to find a standard tool yet. The findings of this research, particularly Figs. 5.12 and 5.13, are still unsatisfactory, but that the statics in many locations keep showing similar characteristics may shed new light on bias correction for the large ensemble of climate simulations.

Acknowledgements This research was supported by the "Integrated Research Program for Advancing Climate Models (TOUGOU program)" of the Ministry of Education, Culture, Sports, Science and Technology, Japan.

References

1. Coles SG (2001) An introduction to statistical modeling of extreme values. Springer, London
2. Hosking JRM, Wallis JR (1997) Regional frequency analysis: an approach based on L-moments. Cambridge University Press, Cambridge

3. IPCC (2014) Climate Change 2014: Synthesis Report. Contribution of working groups I, II and III to the fifth assessment report of the intergovernmental panel on climate change [Core Writing Team, Pachauri RK, Meyer LA (eds)]. IPCC, Geneva, Switzerland. https://www.ipcc.ch/report/ar5/syr/. Cited 8 Dec 2019
4. Iwai S, Ishiguro M (1970) Applied hydrological statistics. Morikita, Tokyo (in Japanese)
5. Japan Meteorological Agency (JMA) (2018) Climate Change Monitoring Report 2017. http://www.jma.go.jp/jma/en/NMHS/ccmr/ccmr2017_low.pdf. Cited 5 Oct 2018
6. Japan Meteorological Agency (JMA) (2020) Download observed meteorological data. http://www.data.jma.go.jp/gmd/risk/obsdl/index.php. Cited 7 Jan 2020
7. Kadoya M (1964) Theory of hydrological statistics. A series of hydraulic engineering, committee on hydraulics, JSCE 64–02:59 (in Japanese)
8. Ministry of Education, Culture, Sports, Science and Technology (MEXT), Japan (2017) Annual report 2016 of program for risk information on climate change, p 319
9. Mizuta R, Murata A, Ishii M, Shiogama H, Hibino K, Mori N, Arakawa O, Imada Y, Yoshida K, Aoyagi Y, Kawase H, Mori M, Okada Y, Shimura T, Nagatomo T, Ikeda M, Endo H, Nosaka M, Arai M, Takahashi C, Tanaka K, Takemi T, Tachikawa Y, Temur K, Kamae Y, Watanabe M, Sasaki H, Kitoh A, Takayabu I, Nakakita E, Kimoto M (2017) Over 5000 years of ensemble future climate simulations by 60 km global and 20 km regional atmospheric models. Bull Amer Meteor Soc July 2017, 1383–1398
10. Reiss RD, Thomas M (1997) Statistical analysis of extreme values. Birkhauser, p 316
11. Seneviratne SI et al (2012) Changes in climate extremes and their impacts on the natural physical environment. In: Field CB et al (eds) Managing the risks of extreme events and disasters to advance climate change adaptation. A special report of working groups I and II of the intergovernmental panel on climate change (IPCC), Cambridge University Press, Cambridge, UK, pp 109–230
12. Stedinger JR, Vogel RM, Foufoula-Georgiou E (1993) Frequency analysis of extreme events. In: Maidment DR (ed) Handbook of hydrology. McGraw-Hill, New York, pp 18.1–18.66
13. UNESCO (1964) Final report of intergovernmental meeting of experts for the IHD. Paris
14. Watanabe S, Kanae S, Seto S, Yeh PJ-F, Hirabayashi Y, Oki T (2012) Intercomparison of bias-correctionmethods for monthly temperature and precipitation simulated by multiple climate models. J Geophys Res 117:D23114. https://doi.org/10.1029/2012JD018192

Chapter 6
History and Perspectives of Hydrologic Frequency Analysis in Japan

Kaoru Takara

Abstract Hydrologic frequency analysis provides basic information for the planning, design, and management of hydraulic and water resources systems for promoting the river basin quality and human health. It uses meteorological/hydrological extreme-value data and probability distribution functions to estimate T-year events (quantiles). Reviewing the history of hydrologic frequency analysis in Japan, this study describes goodness-of-fit criteria such as the standard least-squares criterion and Akaike information criterion and their applications. The jackknife and bootstrap methods are introduced as useful resampling methods for bias correction and quantile variability estimation. As future directions, this study proposes the incorporation of (1) partial duration series or peaks-over-threshold series, if available, instead of the annual maximum series; (2) a nonparametric method using empirical distributions for larger samples with more than 100-year observation period; and (3) probable maximum precipitation or probable maximum flood into frequency analysis.

Keywords Bootstrap · Extreme values · Flood · Goodness-of-fit · Jackknife · Nonparametric · Precipitation · Probable maximum values · Return period

6.1 Introduction

In hydrological research, significant effort has been aimed toward analyzing flood hazards and establishing countermeasures for predicting extreme weather/flood events and coping with resultant flood disasters in river basins. River engineers and managers are interested in estimating reasonable flood discharges to improve the design and planning of flood control and water resources management. Before the implementation of systematic and continuous meteorological and hydrological

K. Takara (✉)
Professor, Graduate School of Advanced Integrated Studies in Human Survivability (Shishu-Kan), Kyoto University, Higashi-Ichijo-Kan, 1 Yoshida-Nakaadachi-Cho, Sakyo-ku, Kyoto 606-8306, Japan
e-mail: takara.kaoru.7v@kyoto-u.ac.jp

N. Hoshino et al. (eds.), *Pioneering Works on Extreme Value Theory*, JSS Research Series in Statistics,
https://doi.org/10.1007/978-981-16-0768-4_6

observations such as rain gauges and water stage observatories, historical maximum events were used for flood control practices. Such historical maxima, however, were often broken by bigger flood events that took place later. Probabilistic analysis of river discharge was initiated in the US in the 1910s [14, 15] and further developed in the 1930s by Gumbel [11].

In Japan, such probabilistic analysis was introduced by Iwai [22] in the 1940s and was officially implemented when a river management manual of the Japanese Ministry of Construction [32] was established, followed by the amendment of a river law in 1964. Hydrologic frequency analysis (HFA) has since played an important role in river engineering, flood control, urban storm drainage, and water resource management in Japan [46, 55]. The manual established in 1958 was revised in 1976. Currently, the Japanese Ministry of Land, Infrastructure, Transport and Tourism translated it into English as "Technical Criteria for River Works" and made various kinds of revisions. Table 6.1 summarizes the history of Japanese frequency analysis research.

6.2 Hydrologic Frequency Analysis Method

Hydrological frequency analysis includes the following steps:

Step 1: Evaluate data homogeneity and its independence.
Step 2: Enumerate several distributions as candidates for quantile estimation.
Step 3: Estimate parameters for each distribution.
Step 4: Screen the distributions to assess goodness of fit.
Step 5: Analyze the variability of quantile estimates for distributions that have not been excluded in the prior step, by using a resampling method such as the jackknife or the bootstrap.
Step 6: Select a distribution that fits data well and exhibits the smallest variability for quantile estimators.

In HFA, we usually use the annual maximum series (AMS) of rainfall or discharge. The annual maxima are basically well recorded in meteorological/hydrological stations in each country. So does Japan. We assume that the data (annual maxima) are independent of each other and follow an identical distribution. Annual second maxima may be larger than annual maxima in other years. It is unclear whether we can ignore such large annual second and third maxima that are larger than annual maxima in other years. Partial duration series (PDS) or peaks-over-threshold (POT) analysis may also be useful for addressing this problem, which is discussed in detail later.

The generalized extreme-value (GEV) distribution is often used for the frequency analysis of hydrological extremes worldwide. In Japan, however, HFA has been developed with normal (Gaussian) distribution theory; the lognormal distribution has been used with lognormal probability paper. The extreme-value distributions were

Table 6.1 History of hydrologic frequency analysis (HFA) in Japan

Publication	Brief notes
Iwai [22]	Introduced overseas research in HFA
Iwai [23]	Analytical solutions of the Slade-type lognormal distribution
Iwai [24]	Applied lognormal distributions to Japanese rivers
Ishihara and Iwai [19]	Proposed to introduce probabilistic methods for river planning
Kadoya [27]	A solution for extreme-value distribution
Ishihara and Takase [21]	Method of moments for lognormal distribution
Takase [54]	Order statistics analysis for the lognormal distribution
Ministry of Const. [32]	Established a manual applying probabilistic methods in practice
Kadoya [28]	Analyzed the lognormal distribution and its parameters
Kadoya [29]	A seminar textbook for hydrological statistics
Iwai and Ishiguro [25]	A textbook "Applied Hydrological Statistics"
Nagao and Kadoya [33]	Two-variate exponential distribution and its numerical table
Hashino [12]	A maximum likelihood method for the lognormal distribution
Kanda [30]*	Proposed a new extreme-value distribution with an upper bound
Kanda and Fujita [31]	A textbook for probabilistic methods in hydrology
Etoh and Murota [6]	Proposed SQRT-k distribution
Hoshi et al. [18]	Monte Carlo experiments for lognormal distribution
Hoshi and Leeyavanija [17]	Introduced the sextile method to Pearson III distributions
Sogawa et al. [38]	Proposed a multivariate maximum entropy distribution
Takasao et al. [53]	Proposed the Standard Least-Square Criterion (SLSC)
Etoh et al. [7, 8]	Proposed and applied SQRT-ET-max (SQET) distribution
Hashino [13]	Rainfall intensity-duration curve for design floods
Takara and Takasao [47, 48]	Applied SLSC and jackknife in HFA
Takara and Loebis [45]	Applied probable maximum events to HFA
Takara et al. [49]	Applied Slade-type distribution with upper and lower bounds
Takara and Tosa [50, 51]	Used distributions with PMP/PMF as upper bound
Tanaka and Takara [56]	Applied SLSC and the jackknife to river discharges in Japan
Tanaka and Takara [57]	Considered AMS and PDS (POT) in flood frequency
Takara et al. [43]	Applied radar data to depth-area-duration (DAD) analysis
Tanaka and Takara [58, 59]	Further considered the compatibility of AMS and PDS
Takara [40, 41]	Nonparametric analysis using the empirical distribution
Ishihara and Nakaegawa [20]	Applied nonparametric analysis of 51 Japanese rain gauges
Takara and Kobayashi [44]	Parametric and nonparametric depending on sample size
Takara [42]	Nonparametric analysis using PMP as an upper bound

*Applied to earthquake motions and wind speeds, not hydrology

also used in parallel as an alternative method. In the United States, Pearson Type III (Gamma) and log-Pearson Type III distributions were also recommended. Then Japanese researchers and practitioners have introduced Pearson Type III distributions as well. Other distributions were also proposed by Japanese researchers, as detailed in the next section. In addition, sometimes the GEV distribution does not fit better with extreme-value data than other distributions. There are many candidate distributions used in Japan; thus, we required criteria to assess and determine the best distribution among them.

6.3 Various Probability Distribution Functions

In Japan, various types of probability distribution functions have been extensively investigated since the 1950s. Takeuchi et al. [55] summarized some distribution functions proposed by Japanese researchers: (1) the SQRT-exponential-type distribution of maxima (SQET) by Etoh et al. [8]; (2) bivariate gamma and binomial distributions by Nagao et al. [33]; and (3) the maximum entropy distribution by Sogawa et al. [38]. This section covers commonly used single-variate probability distribution functions such as lognormal and extreme-value distributions and SQET.

Lognormal distribution: The lognormal distribution with three parameters is described as a probability density function:

$$f(x) = \frac{1}{(x-a)\sigma_y\sqrt{2\pi}} \exp\left[-\frac{1}{2}\left\{\frac{\ln(x-a)-\mu_y}{\sigma_y}\right\}^2\right], \tag{6.1}$$

where x is a hydrologic variate, and a, μ_y, and σ_y are parameters; its reduced variate and transformed variate are, respectively, given as follows:

$$s = \frac{\ln(x-a)-\mu_y}{\sigma_y} \quad \text{and} \quad y = \ln(x-a). \tag{6.2}$$

GEV distribution: Jenkinson [26] found that variates that follow the maximum value distribution can be expressed as a unified cumulative distribution function as follows:

$$F(x) = \begin{cases} \exp[-\{1-k(x-x_0)/\alpha\}^{1/k}], & k \neq 0, \\ \exp[-\exp\{-(x-x_0)/\alpha\}], & k = 0, \end{cases} \tag{6.3}$$

where k, α, and x_0 are parameters of this distribution.

This is called the GEV distribution. Since the Natural Environment Research Council adopted this distribution to annual maximum daily discharges in rivers in the UK, this distribution plays an important role in the UK. Prescott and Walden [36] applied it to extreme values of sea water levels in the UK. Arnell et al. [2] proposed an unbiased plotting position formula for the GEV distribution.

If the parameter k equals 0, the GEV distribution is equivalent to the Gumbel distribution (type I extreme-value distribution for maximum values). If $k < 0\,(k > 0)$, it is called a type II distribution (type III distribution). If $k \neq 0$, $F(x)$ can be written as

$$F(x) = \exp\{-\exp(-s)\}, \tag{6.4}$$

where

$$s = -\frac{1}{k}\ln\left[-\left\{x - \left(x_0 + \frac{\alpha}{k}\right)\right\}/\left(\frac{\alpha}{k}\right)\right]. \tag{6.5}$$

Kadoya [27, 29] called it a log-extreme-value distribution type A and type B when $k < 0, k > 0$, respectively.

SQRT-ET-max distribution (SQET): Etoh et al. [8] proposed the following equation based on their analysis considering rainfall intensity during a storm event.

$$F(x) = \begin{cases} 0, & x < 0, \\ \exp[-\lambda(1 + \sqrt{\beta x})\exp(-\sqrt{\beta x})], & x \geq 0, \end{cases} \tag{6.6}$$

where β and λ are parameters.

6.4 Goodness of Fit

To select the best probability distribution function, we usually evaluate the goodness of fit of various distribution functions to extreme-value datasets. Takara and Takasao [47] introduced four goodness-of-fit criteria. See also Takara and Stedinger [46].

Suppose that s is the reduced or standardized variate for x: $s = g(x)$. Let q be the non-exceedance probability. For q^*, a specific value of q, define s^* as

$$s^* = g(F^{-1}(q^*)), \tag{6.7}$$

where $q = F(x)$ and F is a cumulative probability distribution function. Let $y_1, \ldots, y_N$ be the order statistics of the original observations $x_1, \ldots, x_N$ (y_1 is the smallest value), and q_i be the non-exceedance probability assigned to y_i. Using the transformation function g above, we obtain

$$s_i = g(y_i) \tag{6.8}$$

and

$$r_i = g(F^{-1}(q_i)). \tag{6.9}$$

Before introducing the following quantitative goodness-of-fit criteria, probability distributions were traditionally evaluated (or screened) by the so-called visual consistency based on graphical analysis, which may have subjective judgment by the analysts.

6.4.1 SLSC

Takasao et al. [53] proposed the Standard Least-Squares Criterion (SLSC) based on probability paper analysis to evaluate the linearity of the data (order statistics) plotted on the probability paper:

$$\mathrm{SLSC} = \sqrt{\xi^2_{\min}} \,/\, |s^*_{1-q} - s^*_q|, \tag{6.10}$$

where s^*_{1-q} and s^*_q are specific values of the reduced variates that correspond to the non-exceedance probability $1 - q$ and q, respectively; $\xi^2_{\min}$ is obtained by minimizing

$$\xi^2 = \frac{1}{N}\sum_{i=1}^{N}(s_i - r_i)^2. \tag{6.11}$$

This minimization corresponds to the so-called least-squares method (or one of the graphical fitting methods) based on a plotting position formula such as

$$q_i = \frac{i - \alpha}{N + 1 - 2\alpha}, \tag{6.12}$$

where α is a constant. Takasao et al. [53] recommend the use of Hazen's formula ($\alpha = 0.5$) to give q_i.

The denominator in Eq. (6.10) was introduced to standardize the square root of $\xi^2_{\min}$. Thus, the SLSC can be used to compare the goodness of fit across distributions. Since most of the meteorological/hydrological annual maximum samples had less than 100 data points, almost all the plots fell within a range of non-exceedance probabilities 0.01 and 0.99. Then, the value q is typically taken as $q = 0.99$.

Smaller SLSC values imply better fits. We can compare the goodness of fit of different probability distribution functions based on their SLSC values. If we use SLSC, we can judge the goodness of fit by the SLSC value itself without plotting data on probability paper or without drawing both the histogram and probability density function. This is an advantage of SLSC. When using the maximum likelihood (ML) method instead of the least-squares method, we substitute ξ^2 obtained by the ML method into Eq. (6.10).

Takasao et al. [53] fitted five two-parameter distributions (the normal, lognormal, exponential, Gumbel, and log-Gumbel distributions) by the least-squares method to samples in the Lake Biwa Basin: the monthly and yearly precipitation and inflow, and the annual maximum m-day precipitations ($m = 1, 2,$ and 3). They concluded that SLSC ≈ 0.02 corresponds to a good fit; if SLSC > 0.03, other distributions should be tried. They also compared six plotting position formulas: the Weibull ($\alpha = 0.0$), Adamowski ($\alpha = 0.25$), Blom ($\alpha = 0.375$), Cunnane ($\alpha = 0.4$), Gringorten ($\alpha = 0.44$), and Hazen ($\alpha = 0.5$) formulas. For the 70-year datasets of annual maximum m-day precipitations ($m = 1$, 2, and 3), Hazen's formula yielded better quantile estimates than the other five formulas for the lognormal and Gumbel distributions (Here, "better" means "nearest to those obtained by the ML method."), so it was recommended for graphical frequency analysis using lognormal and Gumbel probability papers and the least-squares method.

6.4.2 COR, MLL, and AIC

Other goodness-of-fit criteria were also considered: the correlation coefficient (COR), maximum log-likelihood (MLL), and Akaike information criterion (AIC).

COR: Write the order statistics of observation as $y_i, i = 1, 2, \ldots, N$. The theoretical quantile of the i-th (plotting) position is denoted by r_i. Then the correlation coefficient (COR) between y_i and r_i is

$$\text{COR} = \frac{\sum_{i=1}^{N}(y_i - \bar{y})(r_i - \bar{r})}{[\{\sum_{i=1}^{N}(y_i - \bar{y})^2\} \cdot \{\sum_{i=1}^{N}(r_i - \bar{r})^2\}]^{1/2}}, \tag{6.13}$$

where $\bar{y}$ and $\bar{r}$ are the means of y_i and r_i, respectively.

The value of COR closer to unity corresponds to a better fit. This probability plot correlation coefficient test has been applied to the normal, lognormal, Gumbel distributions [61], and the Pearson Type III distribution [62]. This approach is often used in ocean wave analysis in Japan [9, 10].

MLL: When fitting a probability distribution function to an extreme-value dataset by the ML method, we usually maximize the log-likelihood function obtained by taking logarithms of the likelihood function owing to computational tractability. For N data $x_1, x_2, \ldots, x_N$, the MLL is given by

$$\text{MLL} = \ln L(\hat{\theta}) = \sum_{i=1}^{N} \ln f(x_i; \hat{\theta}), \tag{6.14}$$

where $f(x; \theta)$ is the probability density function and $\hat{\theta}$ is the ML estimator of the parameter vector θ. When several distributions are fitted to a sample, the distribution that gives the greatest MLL value can be regarded as fitting the best. The MLL is not only the maximum of the log-likelihood; it has some interpretation from the viewpoint of information theory. If the population distribution is known, then the Kullback–Leibler (KL) information is used as an evaluation criterion for the models that approximate the population distribution. In general, it is unknown, then the MLL can be used as an alternative criterion instead of the KL information [37].

AIC: In general, distributions with three free parameters fit better than those with two free parameters. As the number of parameters increases, goodness of fit should appear to improve: the SLSC values decrease and the COR and MLL values increase. Consequently, as long as the SLSC, COR, or MLL is used, distributions having more parameters tend to be evaluated as "better." In the evaluation of models, we must consider model simplicity as well as the goodness of fit. The AIC proposed by Akaike [1] balances the number of parameters and the quality of fit using

$$\mathrm{AIC} = -2\mathrm{MLL} + 2N_p, \tag{6.15}$$

where MLL is the maximum log-likelihood and N_p is the number of free parameters. As N_p increases, the second term of Eq. (6.15) increases, while the first term decreases because the goodness of fit improves (the MLL increases). Akaike [1] suggested that the model that minimizes the AIC is best. The AIC has been effectively applied to hydrological research, for example, in the determination of the optimal order of time-series models [16] and in the evaluation of runoff models [52].

6.4.3 Comparison of Goodness-of-Fit Criteria

Takara and Takasao [48] compared these goodness-of-fit criteria values for several extreme-value datasets. Table 6.2 shows four goodness-of-fit criteria for annual maximum daily precipitation in Osaka, Japan, given in Kanda and Fujita [31]. The values of SLSC, COR, MLL, and AIC are given to 11 distributions.

Table 6.2 Comparison of goodness-of-fit criteria of distributions fitted by the ML method for annual maximum daily precipitation at Osaka (1889–1980)

Distribution	N_p	SLSC	COR	MLL	AIC
Normal	2	0.07937	0.9312	−450.15	904.30
Lognormal	2	0.02996	0.9902	−434.91	873.83
Lognormal	3	0.01666***	0.9970***	−432.82***	871.64**
Pearson III	2	0.06116	0.9685	−438.17	880.34
Log-Pearson III	3	0.01749**	0.9967**	−432.91*	871.82
Pearson III	3	0.03765	0.9865	−432.90**	871.80*
SQET	2	0.02423	0.9932	−433.09	870.18***
Gumel	2	0.04769	0.9846	−434.41	872.83
GEV	3	0.02124	0.9944	−433.17	872.34
Log-Gumel	2	0.03496	0.9895	−434.53	873.06
Log-Gumel	3	0.01858*	0.9960*	−433.17	872.34

N_p: Number of parameters; *** indicates the best for each criterion;
** and * the second and the third, respectively

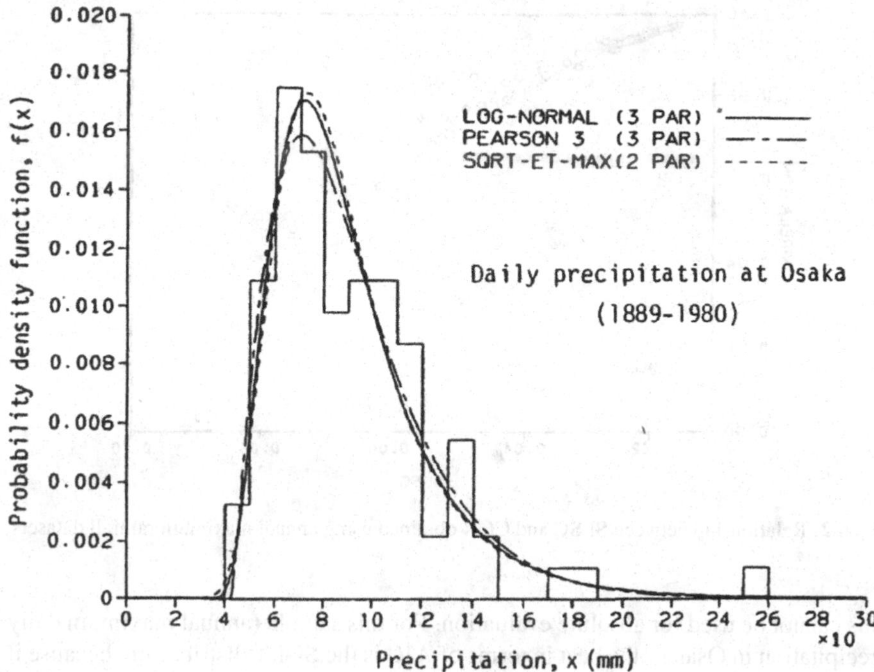

Fig. 6.1 Three distributions fitted by the ML method to the annual maximum daily precipitations at Osaka (1889–1980)

Based on Table 6.2, the three-parameter lognormal distribution is the best for the SLSC, COR, and MLL, while the SQET distribution, which has two parameters, is the best in terms of AIC. The goodness-of-fit criteria can indicate better distributions based on their values. Figure 6.1 shows the histogram of annual maximum daily precipitation at Osaka and the probability distribution functions of three distributions (lognormal, Pearson Type III, and SRET) that indicated better fits in Table 6.2. We could not identify which distribution was the best with these criteria and with visual evaluation such as given in Fig. 6.1.

Takara and Takasao [48] verified an almost one-to-one correspondence between SLSC and COR using the ML method, as shown in Fig. 6.2. SLSC ≈ 0.02 corresponds to COR ≈ 0.995, and SLSC >0.03 corresponds to COR < 0.990. Since the SLSC and COR criteria give almost the same rankings, COR is to be recommended because it is simpler than SLSC. However, SLSC provides a direct understanding that SLSC $= 0.02$ which means approximately 2% error is there between plots and a theoretical line on a probability paper. Therefore, SLSC is often used as a standard method in Japan.

MLL and AIC exhibit the following disadvantages: (1) They are dependent on the ML estimation method, which yields better parameter estimates for larger samples. (2) Their values can be used for comparison, but the values do not have meanings and

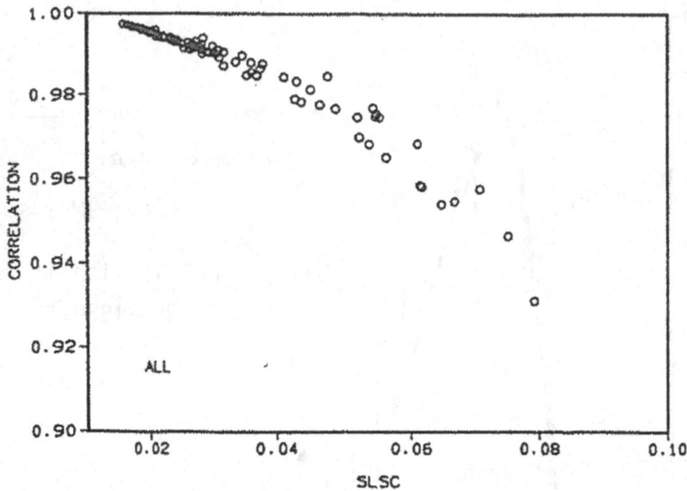

Fig. 6.2 Relationship between SLSC and COR obtained using annual maximum rainfall datasets

thus cannot be used for absolute evaluation. For this sample (annual maximum daily precipitation in Osaka), the best in terms of AIC is the SQET distribution, because it has only two free parameters. AIC has an advantage in that it evaluates simpler (less parameter) distributions. SQET is the fifth in SLSC, the fifth in COR, and the fourth in MLL.

The Japanese river law was amended in 1997. Under this new law, it is necessary for the government to explain the reason for selecting probability distributions to estimate 100-year rainfall and decision-making of design floods based on it. To explain this, they required objective criteria for selecting probability distributions and determining 100-year floods for river planning. Then, the SLSC was adopted as an objective criterion that could be easily applied in practice and easily explain the goodness of fit of the probability distributions considered. SLSC is currently used in almost all river basins in Japan.

There are many types of parameter estimation methods that fit the probability distribution functions to extreme-value datasets. Takara and Stedinger [46] reviewed Monte Carlo experiments reported in the literature, discussing the evaluation of fitting methods for the lognormal (LN), Pearson Type III, log-Pearson Type III, Gumbel, and GEV distributions. Most of the Monte Carlo experiment papers recommended the ML estimation method for two-parameter distributions such as LN(2) and Gumbel distributions. For most of the three-parameter distributions, a combination of the lower bound estimator and the method of moments is useful. For the GEV distribution, the probability weighted moment or L-moments method is useful, see Takara and Stedinger [46] for more details.

6.5 Resampling Methods for Bias Correction and Stability Analysis

Resampling methods such as the jackknife and the bootstrap methods are useful for quantifying the variability (estimation error) of quantile estimators. Resampling methods correct the bias of the statistics obtained from the original dataset and estimate statistical variance, producing many datasets by repeatedly sampling a part of the data from the original set or by repeatedly drawing samples of the same size as the original set with replacement [5]. The datasets produced in this way are relatively easily generated using computers. Examples of these resampling methods are provided by Bardsley [3], Tung and Mays [60], Cover and Unny [4], and Potter and Lettenmaier [34].

A cumulative distribution function $F(x)$ is given by the integration of a density function f with a parameter vector θ:

$$F(x) = \int_{-\infty}^{x} f(t;\theta)dt. \tag{6.16}$$

Using F, we obtain the quantile x_q for a non-exceedance probability q:

$$x_q = F^{-1}(q), \tag{6.17}$$

where the return period (recurrence interval) T years for annual maxima is

$$T = \frac{1}{1-q}. \tag{6.18}$$

Focusing on the quantile x_q (T-year event) in the jackknife (or bootstrap) algorithm, we can easily quantify the variability of quantile estimators. The proposed procedure for selecting a quantile estimation method in **Step 5** compares the variability of the quantile estimate for each distribution obtained by this method.

For the same annual maximum daily precipitation at Osaka above, Table 6.3 shows the jackknife estimates and standard error for three T-year events ($T = 50$, 100, and 200) with regard to distributions selected by the goodness-of-fit criteria. Table 6.3 indicates the following:

(1) The SQET distribution was ranked as the best for the dataset of annual maximum precipitation in Osaka in terms of the smallest variation of the quantile estimates.
(2) The variations (standard errors) of the T-year precipitations for $T = 50$, 100, and 200 for the SQET distribution were about 6.2, 6.7, and 7.1 %, respectively.
(3) Three-parameter distributions tend to give larger variability of quantile estimates. In general, three-parameter distributions fit each dataset well, but larger quantile variation in a penalty results from such flexibility.

Although the maximum likelihood values obtained by the ML method can compare different distributions, the values cannot indicate absolute goodness of fit. Con-

Table 6.3 The jackknife estimates and standard error for T-year daily precipitation at Osaka (in mm, $T = 50$, 100, and 200)

Distribution	N_p	$T = 50$	$T = 100$	$T = 200$
SQET	2	180.46 [11.25]	203.56 [13.58]	227.85 [16.08]
Pearson III	3	172.82 [14.68]	189.34 [17.34]	205.51 [20.36]
Lognormal	3	179.94 [17.83]	201.66 [23.69]	224.17 [30.44]
Log-Pearson III	3	181.99 [19.56]	205.69 [27.06]	230.87 [36.13]
Log-Gumbel	3	182.95 [21.07]	207.62 [30.35]	233.86 [43.03]
GEV	3	183.10 [21.19]	207.83 [30.42]	234.15 [42.13]

The value in the square brackets is the standard error obtained by the jackknife

versely, SLSC and COR can indicate the absolute goodness of fit by the values SLSC < 0.03 or COR > 0.995, although these values are dependent on the plotting position formula used. The SQET distribution, which was ranked as the best in terms of AIC in Table 6.2, showed superior performance in terms of the T-year quantile variability, as shown in Table 6.3. We can verify that it was not ranked within the best three in terms of SLSC in Table 6.2 but showed a good result (SLSC $= 0.02423$).

Tanaka and Takara [56] applied SLSC to annual maximum discharge data at 99 locations (gauging stations) in the 64 major rivers in Japan. They fitted several frequency analysis models (distribution functions) and evaluated the goodness of fit using SLSC. The stability of quantile estimates is also assessed in terms of estimation error obtained by jackknife resampling, which can be used for bias correction and the quantification of estimation error. Based on the application results for 99 samples of annual maximum river discharges, they revealed (1) the goodness of fit of various models; (2) that SLSC $= 0.04$ could be a threshold for well-fitted models for river discharges; and (3) that the jackknife estimation error can be used as an index to evaluate model stability.

6.6 Future Perspectives

6.6.1 AMS or PDS

The frequency analysis method usually uses the AMS. Tanaka and Takara [57] first dealt with the partial duration series (PDS) in Japan, instead of the AMS. The PDS is also called the POT series. The theoretical relationship between AMS and PDS and their analytical methods is described in the Handbook of Hydrology [39].

For the time-series data that exceed a threshold value x_0, the generalized Pareto (GP) distribution can be used. The GP's cumulative distribution function $G(x)$ is defined as follows:

$$G(x) = 1 - \left[1 - \kappa\left(\frac{x - x_0}{a}\right)\right]^{1/\kappa}, \quad \kappa \neq 0, \tag{6.19}$$

where κ, α, and x_0 are parameters of this distribution. If $\kappa = 0$, Eq. (6.19) is simplified as an exponential distribution:

$$G(x) = 1 - \exp\left[-\frac{x - x_0}{a}\right]. \tag{6.20}$$

The Gumbel and GEV distributions are basically used for the AMS, while the exponential and GP distributions are for the PDS. It is known that the GEV shape parameter k in Eq. (6.3) is theoretically compatible with the GP's shape parameter κ in Eq. (6.19) such that k and κ can take the same value.

Tanaka and Takara [57] compared quantile estimates for the AMS and PDS at 79 gauging stations where river discharge is observed in A-class rivers in Japan. They used many traditional probability distributions such as the Gumbel, GEV, and Pearson Type III distributions. They verified the similarity of the AMS-based 100-year quantile estimated by the traditional distributions and PDS-based quantile estimated by the GP distribution. They also confirmed that the jackknife performs well in correcting bias and quantifying the estimation error.

Tanaka and Takara [58] applied the same method to flood peak discharges at 105 locations in A-class rivers in Japan. The average record length was 42 years (maximum length of 67 years and minimum length of 27 years). They focused on the compatibility between the GP (exponential) and GEV (Gumbel) distributions and the shape parameters κ and k. Among the 105 datasets, 17 had similar values of κ and k. For these 17 datasets, the same values of κ and k are obtained, and the quantiles estimated by both are similar to each other.

The other 88 datasets did not exhibit this compatibility. Figure 6.3 shows one of the results that did not indicate compatibility as an example that obtained different values of $\kappa(= -0.017)$ and $k(= -0.209)$. As indicated in Fig. 6.3, the GEV follows the historical maximum value, while GP keeps almost the straight line, which results in the difference between 100-year quantiles obtained by each distribution fitted. The SLSC values for these are also given in the figure (GEV: SLSC = 0.029, GP: SLSC = 0.037), both of which indicate good fit with the river discharge.

Using the exponential and GP distributions for the POT datasets, Tanaka and Takara [59] examined several indices to identify the best method for determining the number of upper extremes best for POT analysis. They revealed the optimum number of upper extremes for 3-, 6-, 12-, and 24-hour rainfalls in two river basins in Japan.

Consequently, through the analyses done in these studies, the PDS is recommended in terms of the shape parameter, quantile estimates, SLSC goodness of fit, and quantile variability. The PDS analysis has the possibility to provide better solutions than the AMS. The data processing to keep such PDS (POT) is necessary in practice.

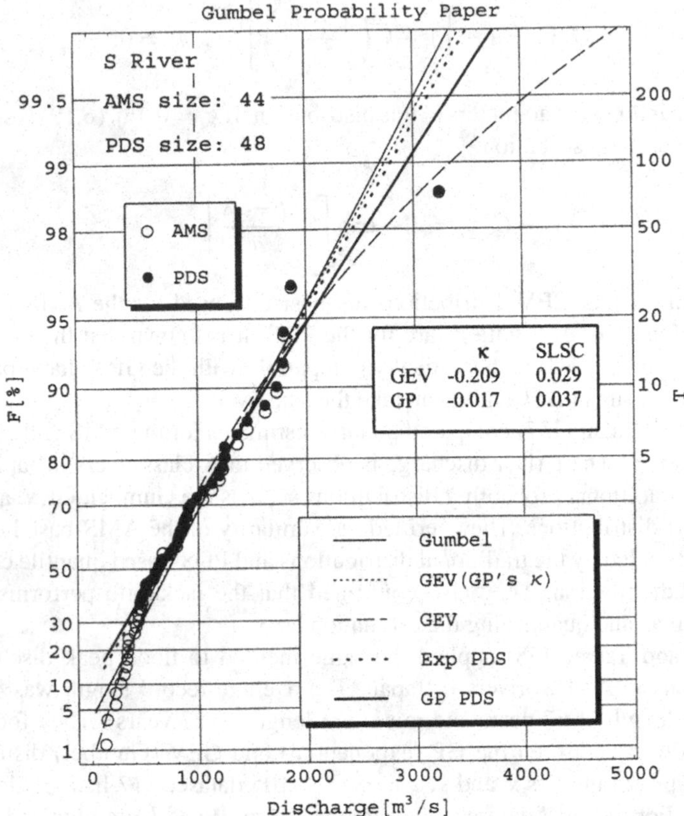

Fig. 6.3 Comparison of GEV and Gumbel distributions fitted to AMS and the GP and exponential distributions fitted to PDS [58]

6.6.2 Nonparametric Analysis for Large Samples

Meteorological and hydrological observations have been developed and modernized systematically, and many data have been accumulated at many observatories of the Japan Meteorological Agency. Fifty years ago, HFA was an extrapolation problem to be solved by using some probability distribution and datasets with records of less than 100 years. Those observatories now have records of more than 100 years. Usually flood control planning is designed to cope with a 100-year flood. This means that estimating 100-year events is now an interpolation problem.

6.6.2.1 Parametric Analysis and Empirical Distribution on the Gumbel Paper

In graphical analysis, we simply connect order statistics as shown in Fig. 6.4. This is an empirical distribution. Takara [40–42], Ishihara and Nakaegawa [20], and Takara and Kobayashi [44] applied such an empirical distribution method to meteorological samples in Japan with a sample size of more than 100.

Figure 6.4 shows an empirical distribution for annual maximum 2-day precipitation in the Ane River Basin. Figure 6.4 also shows the straight solid line, which is a Gumbel distribution fitted by the least-squares method. Focusing on the non-exceedance probability 0.990, which corresponds to the 100-year return period, we obtain 440 mm 2-day precipitation by the empirical distribution method, while we obtain 360 mm by the Gumbel distribution by graphical analysis (the least-squares method). If we apply the Gumbel and GEV distributions fitted by the L-moment

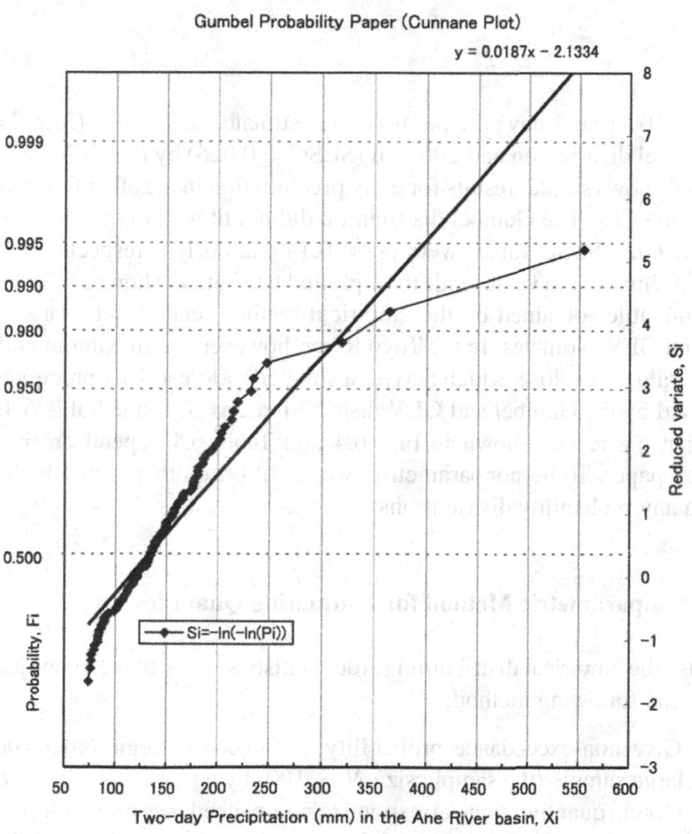

Fig. 6.4 Empirical distribution for annual maximum 2-day precipitations in the Ane River Basin for 108 years (the straight solid line is Gumbel distribution fitted by the least-squares method)

Table 6.4 100-year precipitations (mm) estimated by the empirical distribution compared with the parametric methods: Gumbel and GEV distributions

Distribution	Gumbel (probability paper)		Gumbel		GEV	
Fitting method	Least-squares	**Empirical**	L-moments	SLSC	L-moments	SLSC
Ane River (2 day)	360	**440**	290	0.033	298	0.026
Amano River (daily)	330	**380**	250	0.032	281	0.021
Seri River (daily)	430	**480**	313	0.149	401	0.039
Yogo River (daily)	260	**285**	195	0.035	177	0.020
Toyo River (daily)	302	**285**	302	0.021	300	0.021
Hikone City (daily)	335	**380**	231	0.181	292	0.044

method, the 100-year 2-day precipitations are estimated as 290 mm (SLSC = 0.033) by the Gumbel distribution and 298 mm (SLSC = 0.026) by the GEV distribution.

Table 6.4 shows similar results for daily precipitation in the other four river basins and in Hikone City. The Gumbel distribution did not fit well to the Seri River Basin and Hikone City; SLSC values were poor: 0.149 and 0.181, respectively. At many locations, the maximum value tends to be plotted far right, as shown in Fig. 6.4. Then, 100-year quantiles obtained by the empirical method tended to be larger than the Gumbel and GEV estimates. In the Toyo River, however, the maximum and second-maximum values are close, which derives a smaller 100-year daily precipitation than that obtained by the Gumbel and GEV distributions, as shown in Table 6.4.

Note that the results shown in Fig. 6.4 and Table 6.4 depend on the Gumbel distribution paper. To be nonparametric, we should use other methods that do not depend on any probability distributions.

6.6.2.2 Nonparametric Method for Estimating Quantiles

We may use the empirical distribution (order statistics) to estimate quantiles (T-year events) by the following method:

Step 1: Give non-exceedance probability q_i to each extreme order statistics x_i from a large sample (the sample size $N > 100$) by a plotting position formula.

Step 2: Obtain quantiles that correspond to non-exceedance probabilities $1 - 1/T$ ($T = 10, 20, \ldots, 100$) by the interpolation between (x_{i-1}, q_{i-1}) and (x_i, q_i).

Step 3: Apply these steps (**Step 1** to **Step 2**) by using the bootstrap method for a number of iterations. Then, obtain the bootstrap estimate and variance of the quantiles. This step corrects bias and quantifies the estimation error.

As a plotting position formula in **Step 1**, we recommend the Cunnane formula $(i-0.4)/(N+0.2)$, which is often used in practice recently. The Weibull formula $i/(N+1)$ is used for comparison in this study, see Eq. (6.5).

In particular, for flood control practices, the estimation of 100-year flood is very important. For a large sample with a size of more than 100, the 100-year estimate $x_{0.99}$ is obtained by the interpolation between (x_{N-1}, q_{N-1}), and (x_N, q_N). If we use linear interpolation for a sample with a size of N $(100 < N < 200)$, the 100-year estimate is given by

$$x_{0.99} = x_N - \frac{q_N - 0.99}{q_N - q_{N-1}}(x_N - x_{N-1}). \tag{6.21}$$

When $N = 108$ as an example case, if we use the Cunnane formula, $q_{N-1} = q_{107} = 0.9852126$ and $q_N = q_{108} = 0.9944547$:

$$x_{0.99} = x_N - \frac{0.9944547 - 0.99}{0.9944547 - 0.9852126}(x_N - x_{N-1}) = 0.518x_N + 0.482x_{N-1}. \tag{6.22}$$

The annual maximum 2-day precipitation in the Ane River Basin is $x_N = 554.5$ mm, while the second-maximum $x_{N-1} = 364.4$ mm. Substituting these into Eq. (6.22), we obtain $x_{0.99} = 463$ mm. Using the Weibull formula, we can obtain 537 mm. These values are the same as those in Table 6.5.

Using this method and the same dataset as in Table 6.4, Takara and Kobayashi [44] obtained the results in Table 6.5. The Weibull plotting formula overestimates 100-year quantiles. T-year quantiles estimated by using the Cunnane plotting formula are bias-corrected by the bootstrap, as shown in Table 6.5. Note that the results of empirical distribution (Table 6.4) are similar to the bootstrap results in Table 6.5. This indicates that the proposed new nonparametric method provides reasonable results.

Ishihara and Nakaegawa [20] followed this nonparametric method and applied it to 51 meteorological observatories in Japan with a sample size of 106 (1901–2006). Comparing it with the traditional parametric method, they revealed that the correlation between nonparametric-based 100-year quantile and parametric-based 100-year quantile was 0.98, and the former gives 3 % larger values. The bootstrap method was also useful for this analysis.

Table 6.5 100-year precipitations (mm) estimated by the nonparametric method with the empirical distribution using different plotting positions and the bootstrap estimates [44]

Method Location	Sample size (years)	Weibull plotting	Cunnane plotting	**Cunnane with bootstrap**
Ane River (2 day)	108	537	463	**426**
Amano River (daily)	107	517	413	**383**
Seri River (daily)	107	735	541	**494**
Yogo River (daily)	106	413	315	**288**
Toyo River (daily)	104	286	284	**280**
Hikone City (daily)	100	560	436	**381**

6.6.3 Probability Distribution Functions with Lower and Upper Bounds

A probability distribution may not have a finite lower bound and/or upper bound. For example, the Gaussian (or normal) and Gumbel distributions have no finite bound. Meanwhile, the three-parameter lognormal and log-Pearson Type III distributions have a finite lower bound and no finite upper bound. All these distributions are mathematically relevant and play important roles in practical problems.

From the perspective of scientific rationality, physical variates such as river discharge and rainfall should take positive values (non-negative lower bounds) and have a finite physical maximum limit as an upper bound. Traditionally, however, the negative lower bound has often been accepted even if it is applied to physical variates that cannot be negative because the lower bound is regarded as a location parameter (a free parameter) used to achieve a better fit to the data.

Takara and Tosa [50, 51] addressed two probability distributions that have lower and upper bounds. One is the extreme-value distribution with lower and upper bounds (EVLUB or EV4) distribution used for earthquake motion and wind speed in architectural engineering [30], and the other is a Slade-type lognormal distribution that was introduced to Japan by Iwai in the 1940s [22].

After the tsunami in East Japan in March 2011, possible large events became one of the main concerns of disaster managers and people. The flood analysis is now being implemented by river managers by using possible maximum rainfall with a return period of 1,000 years. The probable maximum precipitation (PMP), probable maximum flood (PMF), and probable maximum tsunami (see Prasad et al. [35]) are now considered by them, although these are not incorporated in frequency analysis practice yet. The author believes the idea to incorporate these probable maximum values, which was proposed in the 1990s by Takara and Loebis [45] and Takara and

Tosa [50, 51], will be used in the future. Such a method can also be linked with the nonparametric method [42].

6.7 Conclusions

This study reviewed the HFA of extreme-value series of precipitations and river discharges in Japan. The review summarized the following:

1. Its history begins with the lognormal distribution applications, because the Gaussian (or normal) distribution was often used in various statistical analyses. The extreme-value distributions are used after it.
2. The goodness-of-fit criteria SLSC, COR, MLL, and AIC were introduced in hydrological frequency analysis. The SLSC is currently a standard goodness-of-fit criteria in practice in Japan.
3. New probability distributions such as SQET and EVLUB were developed by Etoh et al. [8] and Kanda [30], respectively.
4. Computer-aided frequency analysis was initiated by Takasao et al. [53] in the middle of 1980s, including the usage of computer graphics, and computer-intensive statistics using the jackknife and bootstrap resampling methods [47]. The resampling methods are useful in frequency analysis to correct bias and quantify the estimation errors of quantile (T-year event) estimates both in parametric and nonparametric methods.

This study also suggested future directions for frequency analysis in Japan:

(1) Tanaka and Takara [58] compared the AMS and the PDS or POT series. The Gumbel and GEV distributions are generally good for the AMS, while the exponential and GP distributions are suited for the PDS. They recommend the PDS analysis if such PDS (POT) data are available.
(2) Extreme-value datasets are growing. The sample size of many sets exceeds 100. For such large samples, the nonparametric method given by Takara and Kobayashi [44] is recommended because it is fitting method free, goodness of fit free, and reasonable in terms of the theory of order statistics. When using this empirical method, the author recommends reviewing the extreme values plotted on the Gumbel probability paper as well as checking the quantile estimates obtained by the Gumbel and GEV distributions and by the empirical distribution method. This is crucial for verifying the results.
(3) After the tsunami in March 2011, possible large events became one of the main concerns of disaster managers and people. The author believes that the idea of incorporating probable maximum values such as PMP and PMF will be used in practice in the future.

Acknowledgements The author is grateful for the continued support of the Institute of Statistical Mathematics (ISM) for organizing a series of Seminars on Extreme Value Theory and Application for many years. He greatly appreciates Prof. Rinya Takahashi, Prof. Masaaki Sibuya, and

Dr. Takaaki Shimura for arranging the seminars every year and the Pioneering Workshop on Extreme Value and Distribution Theories in honor of Prof. Sibuya held at the ISM on March 21–23, 2019.

References

1. Akaike H (1974) A new look at the statistical model identification. In: IEEE transactions on automatic control, AC-19, pp 716–723
2. Arnell NW, Beran M, Hosking JRM (1986) Unbiased plotting positions for the general extreme value distribution. J Hydrol 86:59–69
3. Bardsley WE (1977) A test for distinguishing between extreme value distributions. J Hydrol 34:377–381
4. Cover KA, Unny TE (1986) Application of computer intensive statistics to parameter uncertainty in streamflow synthesis. Water Resour Res 22(3):495–507
5. Efron B (1982) The Jackknife, the bootstrap and other resampling plans, SIAM monograph, No 38, 92 pp
6. Etoh T, Murota A (1984) A probabilistic model of rainfall of a single storm. J Hydrosci Hydraul Eng, JSCE 4-I:65–77
7. Etoh T, Murota A, Nakanishi M (1987) SQRT-exponential type distribution of maxima, hydrologic frequency modeling. D. Reidel Publishing Co. 253–264
8. Etoh T, Murota A, Yonetani T, Kinoshita T (1986) Frequency of heavy rains. Proceedings Japan society of civil engineers, JSCE, No 369/II-5, pp 165-174 (in Japanese)
9. Goda Y (1988) Numerical analysis on plotting position formulas and confidence interval of quantile estimates in extreme-value statistics. Port Harb Lab Rep 27(1):31–92 (in Japanese)
10. Goda Y (1989) Discussion on "Criteria for evaluating probability distribution models in hydrologic frequency analysis". Takara K, Takasao T (eds) Proceedings Japan society of civil engineers, JSCE, 405 (II-11), pp 265–267 (in Japanese)
11. Gumbel EJ (1958) Statistics of extremes. Columbia University Press, p 375
12. Hashino M (1976) Practical parameter estimation for lognormal distribution by the maximum likelihood method and its application example. Annu J Hydraul Eng, JSCE 20:29–34 (in Japanese)
13. Hashino M (1987) Stochastic formulation of storm pattern and rainfall intensity-duration curve for design flood, hydrologic frequency modeling. D. Reidel Publishing Co. 303–314
14. Hazen A (1914) Storage to be provided in impounding reservoirs for municipal water supply. Trans ASCE 77:1539–1669
15. Hazen A (1930) Flood flows–a study of frequencies and magnitudes. Weiley, New York, p 199
16. Hipel KW, McLead AI, Lennox WC (1977) Advances in box-jenkins modeling–1, model construction. Water Resour Res 13(3):567–575
17. Hoshi K, Leeyavanija U (1986) A new approach to parameter estimations of gamma-type distributions. J Hydrosci Hydraul Eng, JSCE 4(2):79–95
18. Hoshi K, Stedinger JR, Burges SJ (1984) Estimation of log-normal quantiles – Monte Carlo results and first-order approximations. J Hydrol 71:1–30
19. Ishihara T, Iwai S (1949) On rationalization of Japanese river planning in terms of hydrological statistics. Trans JSCE 34(4):24–29 (in Japanese)
20. Ishihara K, Nakaegawa T (2008) Estimation of probable daily precipitation with nonparametric method at 51 meteorological observatories in Japan. J Jpn Soc Hydrol Water Resour 21(6):459–463 (in Japanese)
21. Ishihara T, Takase N (1957) The logarithmic-normal distribution and its solution based on moment method. In: Proceedings of Japan society of civil engineers. No 47, pp 18–23 (in Japanese)
22. Iwai S (1947) On asymmetric distribution in hydrology. In: Proceedings of Japan society of civil engineers, Nos 1–2 (combined issue), pp 93–116 (in Japanese)

23. Iwai S (1949) Examining Slade-type asymmetric distribution and a couple of analytical methods for it. In: Proceedings of Japan society of civil engineers. No 4, pp 84–104 (in Japanese)
24. Iwai S (1949) A method of estimating flood probability and its application to Japanese rivers. Toukei-Suuri-Kenkyuu (Stat Math Res) 2:21–36 (in Japanese)
25. Iwai S, Ishiguro M (1970) Applied hydrological statistics, Morikita Pub., 370 pp (in Japanese)
26. Jenkinson AF (1955) The frequency distribution of the annual maximum (or minimum) of meteorological elements. Q J Roy Meteorol Soc 81(348):158–171
27. Kadoya M (1956) Extreme-value distribution and its solution method. Agric Eng Res 23(6):350–357 (in Japanese)
28. Kadoya M (1962) On the applicable ranges and parameters of logarithmic normal distributions of the Slade-type. Nougyou Doboku Kenkyu (Agricultural Engineering Research). Extra Publication 3:12–27 (in Japanese)
29. Kadoya M (1964) Theory of hydrological statistics. A series of hydraulic engineering, committee on hydraulics, JSCE, 64-02, 59 pp (in Japanese)
30. Kanda J (1981) A new extreme value distribution with lower and upper limits for earthquake motions and wind speeds. Theor Appl Mech. University of Tokyo Press 31:351–354
31. Kanda T, Fujita M (1982) Hydrology – stochastic methods and their applications. Gihodo-Shuppan, Tokyo, p 275
32. Ministry of Const (1958) Manual of technical standards for planning and design of river and sabo engineering (Draft) (in Japanese). Note that this is currently called the "Technical criteria for river works" in English and managed by the Ministry of Land, Infrastructure, Transport and Tourism
33. Nagao M, Kadoya M (1971) Two-variate exponential distribution and its numerical table for engineering application. Bull Disaster Prev Res Inst, Kyoto Univ 20:183–215
34. Potter KW, Lettenmaier DP (1990) A comparison of regional flood frequency estimation methods using a resampling method. Water Resour Res 26(3):415–424
35. Prasad R, Cunningham E, Bagchi G (2009) It Tsunami hazard assessment at nuclear power plant sites in the United States of America - final report (NUREG/CR-6966), USNRC
36. Prescott P, Walden AT (1983) Maximum likelihood estimation of the parameters of the three-parameter generalized extreme-value distribution from censored samples. J Stat Comput Simul 16:241–250
37. Sakamoto H, Ishiguro M, Kitagawa G (1983) Information statistics, Kyoritsu,Tokyo, pp. 27–64 (in Japanese)
38. Sogawa N, Araki A, Sato K (1986) A study on multivariate maximum entropy distribution and its basic characteristics. Proc. JSCE, No 375/II-6, pp 89–98 (in Japanese)
39. Stedinger JR, Vogel RM, Foufoula-Georgiou E (1993) Frequency analysis of extreme events. In: Maidment DR (ed) Handbook of hydrology. McGraw-Hill, 66 pp
40. Takara K (2005) Hydrologic frequency analysis for samples with more than return periods, Recent characteristics of water-related disasters and countermeasures coping with them and hydrologic frequency analysis in an era of large samples, ISM cooperation research seminar "Application of extreme-value theory to engineering," Institute of statistical mathematics, Tokyo, September 8, 2005 (presentation in Japanese)
41. Takara K (2006) Frequency analysis of larger samples of hydrologic extreme-value data — How to estimate the T-year quantile for samples with a size of more than the return period T. In: Annuals, disaster prevention research institute, Kyoto University, No 49 B, pp 7–12 (in Japanese)
42. Takara K (2018) How to incorporate PMP into nonparametric frequency analysis. In: 15th AOGS annual meeting, Honolulu, Hawaii, USA, AS29-A051, June 3–8, 2018
43. Takara K, Hashino T, Nakao T (2001) Application of radar raingauge and nonlinear optimization to DAD analysis. J Hydraul Coast Environ Eng, JSCE 57:1–11 (in Japanese)
44. Takara K, Kobayashi K (2009) Hydrological frequency analysis methods suitable for the sample size of extreme events. Annu J Hydraul Eng, JSCE, pp 205–210 (in Japanese)

45. Takara K, Loebis J (1996) Frequency analysis introducing probable maximum hydrologic events: preliminary studies in Japan and in Indonesia. In: Proceedings of international symposium on comparative research on hydrology and water resources in Southeast Asia and the Pacific, Yogyakarta, Indonesia, 18–22 November 1996, Indonesian National Committee for UNESCO International Hydrology Programme, pp 67–76
46. Takara K, Stedinger J (1994) Recent Japanese contribution to frequency analysis and quantile lower bound estimator. In: Hipel KW (ed) Stochastic and statistical methods in hydrology and environmental engineering: extreme values: floods and droughts, vol 1. Kluwer, Dordrecht, The Netherlands, pp 217–234
47. Takara K, Takasao T (1988) Criteria for evaluating probability distribution models in hydrologic frequency analysis. Proc Jpn Soc Civ Engrs, JSCE, 393 (II-9):151–160 (in Japanese)
48. Takara K, Takasao T (1989) Closure to the discussion by Y. Goda on "Criteria for evaluating probability distribution models in hydrologic frequency analysis". Proc Jpn Soc Civ Engrs, JSCE, 405 (II-11):265–272 (in Japanese)
49. Takara K, Takasao T, Tomosugi K (1996) Possibility and necessity of paradigm shift in hydrologic frequency analysis. In: Proceedings of international conference on water resources and environmental research, Heian-Kaikan, Kyoto, Japan, vol 1, pp 435–442
50. Takara K, Tosa K (1999) Application of probability distributions with lower and upper bounds to hydrologic frequency analysis. Annu J Hydraul Eng, JSCE 43:121–126 (in Japanese)
51. Takara K, Tosa K (1999) Storm and flood frequency analysis using PMP/PMF estimates. In: Proceedings of international symposium on floods and droughts, Nanjing, China, Chinese national committee for UNESCO-IHP 18–20:7–17
52. Takasao T, Shiiba M, Takara K (1984) Evaluation of runoff models by an information criterion. In: Annuals of Disaster Prevention Research Institute, Kyoto University. No 27B-2, pp 275–290 (in Japanese)
53. Takasao T, Takara K, Shimizu A (1986) Basic statistical analysis of hydrological data in lake Biwa Basin. In: Annuals of disaster prevention research institute, Kyoto University. No 29B-2, pp 157–171 (in Japanese)
54. Takase N (1957) Order statistics viewpoints on the logarithmic-normal distribution. In: Proceedings of Japan Society of Civil Engineers. No 47, pp 24–29 (in Japanese)
55. Takeuchi K, Takara K, Etoh T, Hashino M, Nagao M, Sogawa N (1993) Hydrological statistics, research and practice of hydraulic engineering in Japan. J Hydrosci Hydraul Eng, JSCE, No. SI-3, 1993, pp 175–193
56. Tanaka S, Takara K (1999) Goodness-of-fit and stability assessment in flood frequency analysis. Annu J Hydraul Eng, JSCE 43:127–132 (in Japanese)
57. Tanaka S, Takara K (1999) Hydrologic frequency analysis with annual maximum series and partial duration series. Annu J Hydraul Eng, JSCE 43:145–150 (in Japanese)
58. Tanaka S, Takara K (2001) Comparison of AMS and PDS in flood frequency analysis. Annu J Hydraul Eng, JSCE 45:205–210 (in Japanese)
59. Tanaka S, Takara K (2002) A study on threshold selection in POT analysis of extreme floods. In: The extremes of the extremes: extraordinary floods (Proceedings of a symposium held at Reykjavik, Iceland, July 2000). IAHS Publ., no 271, pp 299–304
60. Tung Y-K, Mays L (1981) Risk models for flood levee design. Water Resour Res 17(4):833–841
61. Vogel RM (1986) The probability plot correlation coefficient test for the normal, lognormal, and Gumbel distribution hypothesis. Water Resour Res 22(4):587–590 (with correction , Water Resour Res 23(10):2013)
62. Vogel RM, McMartin DE (1991) Probability plot goodness-of-fit and skewness estimation procedures for the Pearson type III distribution. Water Resour Res 27(12):3149–3158

Printed in the United States
by Baker & Taylor Publisher Services

Printed in the United States
by Baker & Taylor Publisher Services